P9-DNR-142

The Medusa
and the Snail

ALSO BY LEWIS THOMAS

The Lives of a Cell

The Medusa

MORE NOTES

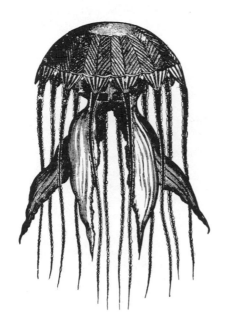

and the Snail

OF A BIOLOGY WATCHER

Lewis Thomas

THE VIKING PRESS

NEW YORK

Copyright © Lewis Thomas, 1974, 1975, 1976, 1977, 1978, 1979
All rights reserved
First published in 1979 by The Viking Press
625 Madison Avenue, New York, N.Y. 10022
Published simultaneously in Canada by
Penguin Books Canada Limited

Library of Congress Cataloging in Publication Data

Thomas, Lewis, 1913–
The medusa and the snail.

1. Biology—Addresses, essays, lectures.
2. Medicine—Addresses, essays, lectures.
I. Title.
QH311.T56   574   79–1199
ISBN 0–670–46568–2

Printed in the United States of America
Set in Videocomp Garamond
Second printing May 1979

Acknowledgments

Daedalus: "Medical Lessons From History" appeared in *Daedalus, Journal of the American Academy of Arts and Sciences*, Boston, Ma., Summer 1977, under the title "Biomedical Science and Human Health: The Long-Range Prospect." Reprinted by permission of *Daedalus*.

Alfred A. Knopf, Inc.: From "The Man with the Blue Guitar," from *The Collected Poems of Wallace Stevens*, Copyright 1936 by Wallace Stevens and renewed 1964 by Holly Stevens. Reprinted by permission of Alfred A. Knopf, Inc.

The New York Times: "The Youngest and Brightest Thing Around" appeared in different form on July 2, 1978, and is reprinted by permission. Copyright © 1978 by The New York Times.

Most of these essays originally appeared in *The New England Journal of Medicine*.

To Beryl, with love

Contents

Contents

The Medusa
and the Snail

The Medusa and the Snail

WE'VE NEVER BEEN so self-conscious about our selves as we seem to be these days. The popular magazines are filled with advice on things to do with a self: how to find it, identify it, nurture it, protect it, even, for special occasions, weekends, how to lose it transiently. There are instructive books, best sellers on self-realization, self-help, self-development. Groups of self-respecting people pay large fees for three-day sessions together, learning self-awareness. Self-enlightenment can be taught in college electives.

You'd think, to read about it, that we'd only just now discovered selves. Having long suspected that there was *something alive* in there, running the place, separate from everything else, absolutely individual and independent, we've celebrated by giving it a real name. My self.

It is an interesting word, formed long ago in much more social ambiguity than you'd expect. The original root was *se* or *seu,* simply the pronoun of the third person, and most of the descendant words, except "self" itself, were constructed to allude to other, somehow connected people; "sibs" and "gossips," relatives and close acquaintances, came from *seu. Se* was also used to indicate something outside or apart, hence words like "separate," "secret," and "segregate." From an extended root *swedh* it moved into Greek as *ethnos,* meaning people of one's own sort, and *ethos,* meaning the customs of such people. "Ethics" means the behavior of people like one's self, one's own ethnics.

We tend to think of our selves as the only wholly unique creations in nature, but it is not so. Uniqueness is so commonplace a property of living things that there is really nothing at all unique about it. A phenomenon can't be unique and universal at the same time. Even individual, free-swimming bacteria can be viewed as unique entities, distinguishable from each other even when they are the progeny of a single clone. Spudich and Koshland have recently reported that motile microorganisms of the same species are like solitary eccentrics in their swimming behavior. When they are searching for food, some tumble in one direction for precisely so many seconds before quitting, while others tumble differently and for different, but characteristic, periods of time. If you watch them closely, tethered by their flagellae to the surface of an antibody-coated slide, you can tell them from each other by the way they twirl, as accurately as though they had different names.

2

Beans carry self-labels, and are marked by these as distinctly as a mouse by his special smell. The labels are glycoproteins, the lectins, and may have something to do with negotiating the intimate and essential attachment between the bean and the nitrogen-fixing bacteria which live as part of the plant's flesh, embedded in root nodules. The lectin from one line of legume has a special affinity for the surfaces of the particular bacteria which colonize that line, but not for bacteria from other types of bean. The system seems designed for the maintenance of exclusive partnerships. Nature is pieced together by little snobberies like this.

Coral polyps are biologically self-conscious. If you place polyps of the same genetic line together, touching each other, they will fuse and become a single polyp, but if the lines are different, one will reject the other.

Fish can tell each other apart as individuals, by the smell of self. So can mice, and here the olfactory discrimination is governed by the same H2 locus which contains the genes for immunologic self-marking.

The only living units that seem to have no sense of privacy at all are the nucleated cells that have been detached from the parent organism and isolated in a laboratory dish. Given the opportunity, under the right conditions, two cells from wildly different sources, a yeast cell, say, and a chicken erythrocyte, will touch, fuse, and the two nuclei will then fuse as well, and the new hybrid cell will now divide into monstrous progeny. Naked cells, lacking self-respect, do not seem to have any sense of self.

The markers of self, and the sensing mechanisms respon-

sible for detecting such markers, are conventionally regarded as mechanisms for maintaining individuality for its
own sake, enabling one kind of creature to defend and
protect itself against all the rest. Selfness, seen thus, is for
self-preservation.

In real life, though, it doesn't seem to work this way. The
self-marking of invertebrate animals in the sea, who must
have perfected the business long before evolution got
around to us, was set up in order to permit creatures of one
kind to locate others, not for predation but to set up symbiotic households. The anemones who live on the shells of
crabs are precisely finicky; so are the crabs. Only a single
species of anemone will find its way to only a single species
of crab. They sense each other exquisitely, and live together
as though made for each other.

Sometimes there is such a mix-up about selfness that
two creatures, each attracted by the molecular configura-
tion of the other, incorporate the two selves to make a
single organism. The best story I've ever heard about
this is the tale told of the nudibranch and medusa living
in the Bay of Naples. When first observed, the nudibranch, a common sea slug, was found to have a tiny
vestigial parasite, in the form of a jellyfish, permanently
affixed to the ventral surface near the mouth. In curiosity
to learn how the medusa got there, some marine biologists began searching the local waters for earlier devel-
opmental forms, and discovered something amazing. The
attached parasite, although apparently so specialized as to
have given up living for itself, can still produce off-
spring, for they are found in abundance at certain sea-

4

sons of the year. They drift through the upper waters, grow up nicely and astonishingly, and finally become full-grown, handsome, normal jellyfish. Meanwhile, the snail produces snail larvae, and these too begin to grow normally, but not for long. While still extremely small, they become entrapped in the tentacles of the medusa and then engulfed within the umbrella-shaped body. At first glance, you'd believe the medusae are now the predators, paying back for earlier humiliations, and the snails the prey. But no. Soon the snails, undigested and insatiable, begin to eat, browsing away first at the radial canals, then the borders of the rim, finally the tentacles, until the jellyfish becomes reduced in substance by being eaten while the snail grows correspondingly in size. At the end, the arrangement is back to the first scene, with the full-grown nudibranch basking, and nothing left of the jellyfish except the round, successfully edited parasite, safely affixed to the skin near the mouth.

It is a confusing tale to sort out, and even more confusing to think about. Both creatures are designed for this encounter, marked as selves so that they can find each other in the waters of the Bay of Naples. The collaboration, if you want to call it that, is entirely specific; it is only this species of medusa and only this kind of nudibranch that can come together and live this way. And, more surprising, they cannot live in any other way; they depend for their survival on each other. They are not really selves, they are specific *others*.

The thought of these creatures gives me an odd feeling. They do not remind me of anything, really. I've never heard of such a cycle before. They are bizarre, that's it, unique. And at the same time, like a vaguely remembered dream, they remind me of the whole earth at once. I cannot get my mind to stay still and think it through.

The Tucson Zoo

SCIENCE GETS MOST of its information by the process of reductionism, exploring the details, then the details of the details, until all the smallest bits of the structure, or the smallest parts of the mechanism, are laid out for counting and scrutiny. Only when this is done can the investigation be extended to encompass the whole organism or the entire system. So we say.

Sometimes it seems that we take a loss, working this way. Much of today's public anxiety about science is the apprehension that we may forever be overlooking the whole by an endless, obsessive preoccupation with the parts. I had a brief, personal experience of this misgiving one afternoon in Tucson, where I had time on my hands and visited the zoo, just outside the city. The designers there have cut a deep pathway between two small artificial ponds, walled by clear glass, so when you stand in the center of the path you can look into the depths of each pool, and at the same time you can regard the surface. In one pool, on the right side of the path, is a family of otters; on the other side, a family of beavers. Within just a few feet from your face, on either side, beavers and otters are at play, underwater and on the surface, swimming toward your face and then away, more filled with life than any creatures I have ever seen before, in all my days. Except for the glass, you could reach across and touch them.

I was transfixed. As I now recall it, there was only one sensation in my head: pure elation mixed with amazement at such perfection. Swept off my feet, I floated from one side to the other, swiveling my brain, staring astounded at the beavers, then at the otters. I could hear shouts across my corpus callosum, from one hemisphere to the other. I remember thinking, with what was left in charge of my consciousness, that I wanted no part of the science of beavers and otters; I wanted never to know how they performed their marvels; I wished for no news about the physiology of their breathing, the coordination of their muscles, their vision, their endocrine systems, their digestive tracts. I hoped never to have to think of them as collections of cells. All I asked for was the full hairy complexity, then in front of my eyes, of whole, intact beavers and otters in motion.

It lasted, I regret to say, for only a few minutes, and then I was back in the late twentieth century, reductionist as ever, wondering about the details by force of habit, but not, this time, the details of otters and beavers. Instead, me. Something worth remembering had happened in my mind, I was certain of that; I would have put it somewhere in the brain stem; maybe this was my limbic system at work. I became a behavioral scientist, an experimental psychologist, an ethologist, and in the instant I lost all the wonder and the sense of being overwhelmed. I was flattened.

But I came away from the zoo with something, a piece of news about myself: I am coded, somehow, for otters and beavers. I exhibit instinctive behavior in their presence, when they are displayed close at hand behind glass, simultaneously below water and at the surface. I have receptors for this display. Beavers and otters possess a "releaser" for me, in the terminology of ethology, and the releasing was my experience. What was released? Behavior. What behavior? Standing, swiveling flabbergasted, feeling exultation and a rush of friendship. I could not, as the result of the transaction, tell you anything more about beavers and otters than you already know. I learned nothing new about them. Only about me, and I suspect also about you, maybe about human beings at large: we are endowed with genes which code out our reaction to beavers and otters, maybe our reaction to each other as well. We are stamped with stereotyped, unalterable patterns of response, ready to be released. And the behavior released in us, by such confrontations, is, essentially, a surprised affection. It is compulsory behavior and we can avoid it only by straining with the full power of our conscious minds, making up conscious excuses all the way. Left to ourselves, mechanistic and autonomic, we hanker for friends.

Everyone says, stay away from ants. They have no lessons for us; they are crazy little instruments, inhuman, incapable of controlling themselves, lacking manners, lacking souls. When they are massed together, all touching, exchanging bits of information held in their jaws like memoranda, they become a single animal. Look out for that. It is a debasement, a loss of individuality, a violation of human nature, an unnatural act.

Sometimes people argue this point of view seriously and with deep thought. Be individuals, solitary and selfish, is the message. Altruism, a jargon word for what used to be called love, is worse than weakness, it is sin, a violation of nature. Be separate. Do not be a social animal. But this is a hard argument to make convincingly when you have to depend on language to make it. You have to print up leaflets or publish books and get them bought and sent around, you have to turn up on television and catch the attention of millions of other human beings all at once, and then you have to say to all of them, all at once, all collected and paying attention: be solitary; do not depend on each other. You can't do this and keep a straight face.

Maybe altruism is our most primitive attribute, out of reach, beyond our control. Or perhaps it is immediately at hand, waiting to be released, disguised now, in our kind of civilization, as affection or friendship or attachment. I don't see why it should be unreasonable for all human beings to have strands of DNA coiled up in chromosomes, coding out instincts for usefulness and helpfulness. Usefulness may turn out to be the hardest test of fitness for survival, more important than aggression, more effective, in the long run, than grabbiness. If this is the sort of information biological science holds for the future, applying to us as well as to ants, then I am all for science.

One thing I'd like to know most of all: when those ants have made the Hill, and are all there, touching and exchanging, and the whole mass begins to behave like a single huge creature, and *thinks,* what on earth is that thought? And while you're at it, I'd like to know a second thing: when it happens, does any single ant know about it? Does his hair stand on end?

The Youngest and Brightest Thing Around

(NOTES FOR A MEDICAL SCHOOL
COMMENCEMENT ADDRESS)

DOCTORS:

Somewhere, on some remote planet set at precisely the right distance from a star of just the right magnitude and the right temperature, on the other side of our galaxy, there is at this moment a committee nearing the end of a year-long study of our own tiny, provincial solar system. The intelligent beings of that place are putting their signatures (numbers of some sort, no doubt) to a paper which asserts, with finality, that life is out of the question here and the place is not worth an expedition. Their instruments have detected the presence of that most lethal of all gases, oxygen, and that is the end of that. They had planned to come, bringing along mobile factories for manufacturing life-giving ammonia, but what's the use of risking strangulation?

The only part of this scenario that I really believe in is that committee. I take it as an article of faith that this is the most fundamental aspect of nature that we know about. If you are going to go looking for evidences of life on other celestial bodies, you need special instruments with delicate sensors for detecting the presence of committees. If there is life there, you will find consortia, collaborating groups, working parties, all over the place.

At least this is true for our kind of life.

Mars, from the look we've had at it thus far, is a horrifying place. It is, by all appearances, stone dead, surely the deadest place any of us has ever seen, hard to look at without flinching. Come to think of it, it is probably the only really dead place of any size we've ever caught a close glimpse of, and the near view is incredibly sad.

Or maybe there is life on Mars, and we've simply missed it so far. The innumerable consultants orbiting around NASA are confounded, just now, by intense arguments, highly technical, over this point. Could there be an island of life at the bottom of one of the Martian ravines? Shouldn't we set down fleets of wheeled vehicles on various parts of the surface, deployed to nose about from place to place, in and out of deep crevices, turning over rocks, sniffing for life? Maybe there is a single spot, just one, where living organisms are holed up.

Maybe so, but if so it would be the strangest thing of all, absolutely incomprehensible. For we are not familiar with this kind of living. We do not have solitary, isolated creatures. It is beyond our imagination to conceive of a single form of life that exists alone and independent, unattached to other forms.

If you dropped a vehicle, or a billion vehicles, for that matter, on our planet you might be able to find one or two lifeless spots, but only if you took very small samples. There are living cells in our hottest deserts and at the tops of our coldest mountains. Even in the ancient frozen rocks recently dug out in Antarctica there are endolithic organisms tucked up comfortably in porous spaces beneath the rock face, as much alive as the petunia in the florist's window.

If you did find a single form of life on Mars, in a single place, how would you go about explaining it? The technical term for this arrangement is a "closed ecosystem," and there is the puzzle. We do not have closed ecosystems here, at all. The only closed ecosystem we know about is the earth itself, and even here the term has to be expanded to include the sun as part of the system, and lord knows what sorts of essential minerals that have drifted onto our surface from outside, at one time or another long ago.

Everything here is alive thanks to the living of everything else. All the forms of life are connected. This is what I meant in proposing the committee as the basis of terrestrial life. The most centrally placed committee, carrying the greatest responsibility, more deeply involved in keeping the whole system running than any other body, or any other working part of the earth's whole body, is the vast community of prokaryotic, nonnucleated, microbes. Without bacteria for starters, we would never have had enough oxygen to go around, nor could we have found and fixed the nitrogen for making enzymes, nor could we recycle the solid matter of life for new generations.

One technical definition of a system is as follows: a system is a structure of interacting, intercommunicating components that, as a group, act or operate individually and jointly to achieve a common goal through the concerted activity of the individual parts. This is, of course, a completely satisfactory definition of the earth, except maybe for that last part about a common goal. What on earth is *our* common goal? How did we ever get mixed up in a place like this?

This is the greatest discomfort for our species. Some of us simply write it off by announcing that our situation is ridiculous, that the whole place is ungovernable, and that our responsibilities are therefore to ourselves alone. And yet, there it is: we are components in a dense, fantastically complicated system of life, we are enmeshed in the interliving, and we really don't know what we're up to.

The earth holds together, its tissues cohere, and it has the look of a structure that really would make comprehensible sense if only we knew enough about it. From a little way off, photographed from the moon, it seems to be a kind of organism. Looked at over its whole time, it is plainly in the process of developing, like an enormous embryo. It is, for all its stupendous size and the numberless units and infinite variety of its life forms, coherent. Every tissue is linked for its viability to every other tissue; it gets along by symbiosis, and the invention of new modes of symbiotic coupling is a fundamental process in its embryogenesis. We have no rules for the evolution of this kind of life. We have learned a lot, and in some biomathematical detail, about the laws governing the evolution of individual species on the earth, but no Darwin has yet emerged to take account of the orderly, coordinated growth and differentiation of the whole astonishing system, much less its seemingly permanent survival. It makes an interesting problem: how do mechanisms that seem to be governed entirely by chance and randomness bring into existence new species which fit so neatly and precisely, and usefully, as though they were the cells of an organism? This is a wonderful puzzle.

And now human beings have swarmed like bees over the whole surface, changing everything, meddling with all the other parts, making believe we are in charge, risking the survival of the entire magnificent creature.

You could forgive us, or excuse us anyway, on grounds of ignorance, and at least it can be said for us that we are, at long last, becoming aware of that. In no other century of our brief existence have human beings learned so deeply, and so painfully, the extent and depth of their ignorance about nature. We are beginning to confront this, and trying to do something about it with science, and this may save us all if we are clever enough, and lucky enough. But we are starting almost from scratch, and we have a long, long way to go.

Mind you, I do not wish to downgrade us; I believe fervently in our species and have no patience with the current fashion of running down the human being as a useful part of nature. On the contrary, we are a spectacular, splendid manifestation of life. We have language and can build metaphors as skillfully and precisely as ribosomes make proteins. We have affection. We have genes for usefulness, and usefulness is about as close to a "common goal" for all of nature as I can guess at. And finally, and perhaps best of all, we have music. Any species capable of producing, at this earliest, juvenile stage of its development—almost instantly after emerging on the earth by any evolutionary standard —the music of Johann Sebastian Bach, cannot be all bad. We ought to be able to feel more secure for our future, with Julian of Norwich at our elbow: "But all shall be well and all shall be well and all manner of thing shall be well." For our times of guilt we have Montaigne to turn to: "If it did not seem crazy to talk to oneself, there is not a day when I would not be heard growling at myself, 'Confounded fool.'"

But security is the last thing we feel entitled to feel. We are, perhaps uniquely among the earth's creatures, the worrying animal. We worry away our lives, fearing the future, discontent with the present, unable to take in the idea of dying, unable to sit still. We deserve a better press, in my view. We have always had a strong hunch about our origin, which does us credit; from the oldest language we know, the Indo-European tongue, we took the word for earth— *Dhghem*—and turned it into "humus" and "human"; "humble" too, which does us more credit. We are by all odds the most persistently and obsessively social of all species, more dependent on each other than the famous social insects, and really, when you look at us, infinitely more imaginative and deft at social living. We are good at this; it is the way we have built all our cultures and the literature of our civilizations. We have high expectations and set high standards for our social behavior, and when we fail at it and endanger the species—as we have done several times in this century—the strongest words we can find to condemn ourselves and our behavior are the telling words "inhuman" and "inhumane."

There is nothing at all absurd about the human condition. We matter. It seems to me a good guess, hazarded by a good many people who have thought about it, that we may be engaged in the formation of something like a mind for the life of this planet. If this is so, we are still at the most primitive stage, still fumbling with language and thinking, but infinitely capacitated for the future. Looked at this way, it is remarkable that we've come as far as we have in so short a period, really no time at all as geologists measure time. We are the newest, the youngest, and the brightest thing around.

On Magic in Medicine

MEDICINE HAS ALWAYS BEEN under pressure to provide public explanations for the diseases with which it deals, and the formulation of comprehensive, unifying theories has been the most ancient and willing preoccupation of the profession. In the earliest days, hostile spirits needing exorcism were the principal pathogens, and the shaman's duty was simply the development of improved techniques for incantation. Later on, especially in the Western world, the idea that the distribution of body fluids among various organs determined the course of all illnesses took hold, and we were in for centuries of bleeding, cupping, sweating, and purging in efforts to intervene. Early in this century the theory of autointoxication evolved, and a large part of therapy was directed at emptying the large intestine and keeping it empty. Then the global concept of focal infection became popular, accompanied by the linked notion of allergy to the presumed microbial pathogens, and no one knows the resulting toll of extracted teeth, tonsils, gallbladders, and appendixes: the idea of psychosomatic influences on disease emerged in the 1930s and, for a while, seemed to sweep the field.

Gradually, one by one, some of our worst diseases have been edited out of such systems by having their causes indisputably identified and dealt with. Tuberculosis was the paradigm. This was the most chronic and inexorably progressive of common human maladies, capable of affecting virtually every organ in the body and obviously influenced by crowding, nutrition, housing, and poverty; theories involving the climate in general, and night air and insufficient sunlight in particular, gave rise to the spa as a therapeutic institution. It was not until the development of today's effective chemotherapy that it became clear to everyone that the disease had a single, dominant, central cause. If you got rid of the tubercle bacillus you were rid of the disease.

But that was some time ago, and today the idea that complicated diseases càn have single causes is again out of fashion. The microbial infections that can be neatly coped with by antibiotics are regarded as lucky anomalies. The new theory is that most of today's human illnesses, the infections aside, are multifactorial in nature, caused by two great arrays of causative mechanisms: 1) the influence of things in the environment and 2) one's personal life-style. For medicine to become effective in dealing with such diseases, it has become common belief that the environment will have to be changed, and personal ways of living will also have to be transformed, and radically.

These things may turn out to be true, for all I know, but it will take a long time to get the necessary proofs. Meanwhile, the field is wide open for magic.

One great difficulty in getting straightforward answers is that so many of the diseases in question have unpredictable courses, and some of them have a substantial tendency toward spontaneous remission. In rheumatoid arthritis, for instance, when such widely disparate therapeutic measures as copper bracelets, a move to Arizona, diets low in sugar or salt or meat or whatever, and even an inspirational book have been accepted by patients as useful, the trouble in evaluation is that approximately 35 percent of patients with this diagnosis are bound to recover no matter what they do. But if you actually have rheumatoid arthritis or, for that matter, schizophrenia, and then get over it, or if you are a doctor and observe this to happen, it is hard to be persuaded that it wasn't *something* you did that was responsible. Hence you need very large numbers of patients and lots of time, and a cool head.

Magic is back again, and in full force. Laetrile cures cancer, acupuncture is useful for deafness and low-back pain, vitamins are good for anything, and meditation, yoga, dancing, biofeedback, and shouting one another down in crowded rooms over weekends are specifics for the human condition. Running, a good thing to be doing for its own sake, has acquired the medicinal value formerly attributed to rare herbs from Indonesia.

There is a recurring advertisement, placed by Blue Cross
on the op-ed page of *The New York Times,* which urges you
to take advantage of science by changing your life habits,
with the suggestion that if you do so, by adopting seven
easy-to-follow items of life-style, you can achieve eleven
added years beyond what you'll get if you don't. Since
today's average figure is around seventy-two for all parties
in both sexes, this might mean going on until at least the age
of eighty-three. You can do this formidable thing, it is
claimed, by simply eating breakfast, exercising regularly,
maintaining normal weight, not smoking cigarettes, not
drinking excessively, sleeping eight hours each night, and
not eating between meals.

The science which produced this illumination was a careful study by California epidemiologists, based on a questionnaire given to about seven thousand people. Five years
after the questionnaire, a body count was made by sorting
through the county death certificates, and the 371 people
who had died were matched up with their answers to the
questions. To be sure, there were more deaths among the
heavy smokers and drinkers, as you might expect from the
known incidence of lung cancer in smokers and cirrhosis
and auto accidents among drinkers. But there was also a
higher mortality among those who said they didn't eat
breakfast, and even higher in those who took no exercise,
no exercise at all, not even going off in the family car for
weekend picnics. Being up to 20 percent overweight was
not so bad, surprisingly, but being *underweight* was clearly
associated with a higher death rate.

The paper describing these observations has been widely

quoted, and not just by Blue Cross. References to the Seven Healthy Life Habits keep turning up in popular magazines and in the health columns of newspapers, always with that promise of eleven more years.

The findings fit nicely with what is becoming folk doctrine about disease. You become ill because of not living right. If you get cancer it is, somehow or other, your own fault. If you didn't cause it by smoking or drinking or eating the wrong things, it came from allowing yourself to persist with the wrong kind of personality, in the wrong environment. If you have a coronary occlusion, you didn't run enough. Or you were too tense, or you *wished* too much, and didn't get a good enough sleep. Or you got fat. Your fault.

But eating breakfast? It is a kind of enchantment, pure magic.

You have to read the report carefully to discover that there is another, more banal way of explaining the findings. Leave aside the higher deaths in heavy smokers and drinkers, for there is no puzzle in either case; these are dangerous things to do. But it is hard to imagine any good reason for dying within five years from not eating a good breakfast, or any sort of breakfast.

The other explanation turns cause and effect around. Among the people in that group of seven thousand who answered that they don't eat breakfast, don't go off on picnics, are underweight, and can't sleep properly, there were surely some who were already ill when the questionnaire arrived. They didn't eat breakfast because they couldn't stand the sight of food. They had lost their appetites, were losing weight, didn't feel up to moving around much, and had trouble sleeping. They didn't play tennis or go off on family picnics because they didn't *feel* good. Some of these people probably had an undetected cancer, perhaps of the pancreas; others may have had hypertension or early kidney failure or some other organic disease which the questionnaire had no way of picking up. The study did not ascertain the causes of death in the 371, but just a few deaths from such undiscerned disorders would have made a significant statistical impact. The author of the paper was careful to note these possible interpretations, although the point was not made strongly, and the general sense you have in reading it is that you can live on and on if only you will eat breakfast and play tennis.

The popular acceptance of the notion of Seven Healthy Life Habits, as a way of staying alive, says something important about today's public attitudes, or at least the attitudes in the public mind, about disease and dying. People have always wanted causes that are simple and easy to comprehend, and about which the individual can *do* something. If you believe that you can ward off the common causes of premature death—cancer, heart disease, and stroke, diseases whose pathogenesis we really do not understand—by jogging, hoping, and eating and sleeping regularly, these are good things to believe even if not necessarily true. Medicine has survived other periods of unifying theory, constructed to explain all of human disease, not always as benign in their effects as this one is likely to be. After all, if people can be induced to give up smoking, stop overdrinking and overeating, and take some sort of regular exercise, most of them are bound to feel the better for leading more orderly, regular lives, and many of them are surely going to look better.

Nobody can say an unfriendly word against the sheer goodness of keeping fit, but we should go carefully with the promises.

There is also a bifurcated ideological appeal contained in the seven-life-habits doctrine, quite apart from the subliminal notion of good luck in the numbers involved (7 come 11). Both ends of the political spectrum can find congenial items. At the further right, it is attractive to hear that the individual, the good old freestanding, free-enterprising American citizen, is responsible for his own health and when things go wrong it is his own damn fault for smoking and drinking and living wrong (and he can jolly well pay for it). On the other hand, at the left, it is nice to be told that all our health problems, including dying, are caused by failure of the community to bring up its members to live properly, and if you really want to improve the health of the people, research is not the answer; you should upheave the present society and invent a better one. At either end, you can't lose.

In between, the skeptics in medicine have a hard time of it. It is much more difficult to be convincing about ignorance concerning disease mechanisms than it is to make claims for full comprehension, especially when the comprehension leads, logically or not, to some sort of action. When it comes to serious illness, the public tends, understandably, to be more skeptical about the skeptics, more willing to believe the true believers. It is medicine's oldest dilemma, not to be settled by candor or by any kind of rhetoric; what it needs is a lot of time and patience, waiting for science to come in, as it has in the past, with the solid facts.

The Wonderful Mistake

THE GREATEST single achievement of nature to date was surely the invention of the molecule of DNA. We have had it from the very beginning, built into the first cell to emerge, membranes and all, somewhere in the soupy water of the cooling planet three thousand million years or so ago. All of today's DNA, strung through all the cells of the earth, is simply an extension and elaboration of that first molecule. In a fundamental sense we cannot claim to have made progress, since the method used for growth and replication is essentially unchanged.

But we have made progress in all kinds of other ways. Although it is out of fashion today to talk of progress in evolution if you use that word to mean anything like improvement, implying some sort of value judgment beyond the reach of science, I cannot think of a better term to describe what has happened. After all, to have come all the way from a system of life possessing only one kind of primitive microbial cell, living out colorless lives in hummocks of algal mats, to what we see around us today—the City of Paris, the State of Iowa, Cambridge University, Woods Hole, the succession of travertine-lined waterfalls and lakes like flights of great stairs in Yugoslavia's Plitvice, the horse-chestnut tree in my backyard, and the columns of neurones arranged in modules in the cerebral cortex of vertebrates—*has* to represent improvement. We have come a long way on that old molecule.

We could never have done it with human intelligence, even if molecular biologists had been flown in by satellite at the beginning, laboratories and all, from some other solar system. We have evolved scientists, to be sure, and so we know a lot about DNA, but if our kind of mind had been confronted with the problem of designing a similar replicating molecule, starting from scratch, we'd never have succeeded. We would have made one fatal mistake: our molecule would have been perfect. Given enough time, we would have figured out how to do this, nucleotides, enzymes, and all, to make flawless, exact copies, but it would never have occurred to us, thinking as we do, that the thing had to be able to make errors.

The capacity to blunder slightly is the real marvel of DNA. Without this special attribute, we would still be anaerobic bacteria and there would be no music. Viewed individually, one by one, each of the mutations that have brought us along represents a random, totally spontaneous accident, but it is no accident at all that mutations occur; the molecule of DNA was ordained from the beginning to make small mistakes.

If we had been doing it, we would have found some way to correct this, and evolution would have been stopped in its tracks. Imagine the consternation of human scientists, successfully engaged in the letter-perfect replication of prokaryotes, nonnucleated cells like bacteria, when nucleated cells suddenly turned up. Think of the agitated commissions assembled to explain the scandalous proliferation of trilobites all over the place, the mass firings, the withdrawal of tenure.

To err is human, we say, but we don't like the idea much, and it is harder still to accept the fact that erring is biological as well. We prefer sticking to the point, and insuring ourselves against change. But there it is: we are here by the purest chance, and by mistake at that. Somewhere along the line, nucleotides were edged apart to let new ones in; maybe viruses moved in, carrying along bits of other, foreign genomes; radiation from the sun or from outer space caused tiny cracks in the molecule, and humanity was conceived.

And maybe, given the fundamental instability of the molecule, it had to turn out this way. After all, if you have a mechanism designed to keep changing the ways of living, and if all the new forms have to fit together as they plainly do, and if every improvised new gene representing an embellishment in an individual is likely to be selected for the species, and if you have enough time, maybe the system is simply bound to develop brains sooner or later, and aware-ness.

Biology needs a better word than "error" for the driving force in evolution. Or maybe "error" will do after all, when you remember that it came from an old root meaning to wander about, looking for something.

Ponds

LARGE AREAS of Manhattan are afloat. I remember when the new Bellevue Hospital was being built, fifteen years ago; the first stage was the most spectacular and satisfying, an enormous square lake. It was there for the two years, named Lake Bellevue, while the disconsolate Budget Bureau went looking for cash to build the next stage. It was fenced about and visible only from the upper windows of the old hospital, but pretty to look at, cool and blue in midsummer, frozen gleaming as Vermont in January. The fence, like all city fences, was always broken, and we could have gone down to the lake and used it, but it was known to be an upwelling of the East River. At Bellevue there were printed rules about the East River: if anyone fell in, it was an emergency for the Infectious-Disease Service, and the first measures, after resuscitation, were massive doses of whatever antibiotics the hospital pharmacy could provide.

But if you cleaned the East River you could have ponds all over town, up and down the East Side of Manhattan anyway. If you lifted out the Empire State Building and the high structures nearby, you would have, instantly, an inland sea. A few holes bored in the right places would let water into the subways, and you'd have lovely underground canals all across to the Hudson, uptown to the Harlem River, downtown to the Battery, a Venice underground, without pigeons.

It wouldn't work, though, unless you could find a way to keep out the fish. New Yorkers cannot put up with live fish out in the open. I cannot explain this, but it is so.

There is a new pond, much smaller than Lake Bellevue, on First Avenue between Seventieth and Seventy-first, on the east side of the street. It emerged sometime last year, soon after a row of old flats had been torn down and the hole dug for a new apartment building. By now it is about average size for Manhattan, a city block long and about forty feet across, maybe eight feet deep at the center, more or less kidney-shaped, rather like an outsized suburban swimming pool except for the things floating, and now the goldfish.

With the goldfish, it is almost detestable. There are, clearly visible from the sidewalk, hundreds of them. The neighborhood people do not walk by and stare into it through the broken fence, as would be normal for any other

Manhattan pond. They tend to cross the street, looking away.

Now there are complaints against the pond, really against the goldfish. How could people do such a thing? Bad enough for pet dogs and cats to be abandoned, but who could be so unfeeling as to abandon goldfish? They must have come down late at night, carrying their bowls, and simply dumped them in. How could they?

The ASPCA was called, and came one afternoon with a rowboat. Nets were used, and fish taken away in new custodial bowls, some to Central Park, others to ASPCA headquarters, to the fish pound. But the goldfish have multiplied, or maybe those people with their bowls keep coming down late at night for their furtive, unfeeling dumping. Anyway, there are too many fish for the ASPCA, for which this seems to be a new kind of problem. An official stated for the press that the owners of the property would be asked to drain the pond by pumping, and then the ASPCA would come back with nets to catch them all.

You'd think they were rats or roaches, the way people began to talk. Get those goldfish out of that pond, I don't care how you do it. Dynamite, if necessary. But get rid of them. Winter is coming, someone said, and it is deep enough so that they'll be swimming around underneath the ice. Get them out.

It is this knowledge of the East River, deep in the minds of all Manhattan residents, more than the goldfish themselves, I think. Goldfish in a glass bowl are harmless to the human mind, maybe even helpful to minds casting about for something, anything, to think about. But goldfish let loose, propagating themselves, worst of all *surviving* in what has to be a sessile eddy of the East River, somehow threaten us all. We do not like to think that life is possible under some conditions, especially the conditions of a Manhattan pond. There are four abandoned tires, any number of broken beer bottles, fourteen shoes and a single sneaker, and a visible layer, all over the surface, of that grayish-green film that settles on all New York surfaces. The mud at the banks of the pond is not proper country mud but reconstituted Manhattan landfill, ancient garbage, fossilized coffee grounds and grapefruit rind, the defecation of a city. For goldfish to be swimming in such water, streaking back and forth mysteriously in small schools, feeding, obviously feeding, looking as healthy and well-off as goldfish in the costliest kind of window-box aquarium, means something is wrong with our standards. It is, in some deep sense beyond words, insulting.

I thought I noticed a peculiar sort of fin on the undersurface of two of the fish. Perhaps, it occurs to me now in a rush of exultation, in such a pond as this, with all its chemical possibilities, there are contained some mutagens, and soon there will be schools of mutant goldfish. Give them just a little more time, I thought. And then, with the most typically Manhattan thought I've ever thought, I thought: The ASPCA will come again, next month, with their rowboat and their nets. The proprietor will begin pumping out the pond. The nets will flail, the rowboat will settle, and then the ASPCA officials will give a sudden shout of great dismay. And with a certain amount of splashing and grayish-greenish spray, at all the edges of the pond, up all the banks of ancient New York landfill mud, crawling on their new little feet, out onto the sidewalks, up and down and across the street, into doorways and up the fire escapes, some of them with little suckers on their little feet, up the sides of buildings and into open windows, looking for something, will come the goldfish.

It won't last, of course. Nothing like this ever does. The mayor will come and condemn it in person. The Health Department will come and recommend the purchase of cats from out of town because of the constitutional boredom of city cats. The NIH will send up teams of professionals from Washington with a new kind of antifish spray, which will be recalled four days later because of toxicity to cats.

After a few weeks it will be finished anyway, like a lot of New York events. The goldfish will dive deep and vanish, the pond will fill up with sneakers, workmen will come and pour concrete over everything, and by next year the new building will be up and occupied by people all unaware of their special environmental impact. But what a time it was.

To Err Is Human

EVERYONE MUST have had at least one personal experience with a computer error by this time. Bank balances are suddenly reported to have jumped from $379 into the millions, appeals for charitable contributions are mailed over and over to people with crazy-sounding names at your address, department stores send the wrong bills, utility companies write that they're turning everything off, that sort of thing. If you manage to get in touch with someone and complain, you then get instantaneously typed, guilty letters from the same computer, saying, "Our computer was in error, and an adjustment is being made in your account."

These are supposed to be the sheerest, blindest accidents. Mistakes are not believed to be part of the normal behavior of a good machine. If things go wrong, it must be a personal, human error, the result of fingering, tampering, a button getting stuck, someone hitting the wrong key. The computer, at its normal best, is infallible.

I wonder whether this can be true. After all, the whole point of computers is that they represent an extension of the human brain, vastly improved upon but nonetheless human, superhuman maybe. A good computer can think clearly and quickly enough to beat you at chess, and some of them have even been programmed to write obscure verse. They can do anything we can do, and more besides.

It is not yet known whether a computer has its own consciousness, and it would be hard to find out about this. When you walk into one of those great halls now built for the huge machines, and stand listening, it is easy to imagine that the faint, distant noises are the sound of thinking, and the turning of the spools gives them the look of wild creatures rolling their eyes in the effort to concentrate, choking with information. But real thinking, and dreaming, are other matters.

On the other hand, the evidences of something like an *unconscious,* equivalent to ours, are all around, in every mail. As extensions of the human brain, they have been constructed with the same property of error, spontaneous, uncontrolled, and rich in possibilities.

Mistakes are at the very base of human thought, embedded there, feeding the structure like root nodules. If we were not provided with the knack of being wrong, we could never get anything useful done. We think our way along by choosing between right and wrong alternatives, and the wrong choices have to be made as frequently as the right ones. We get along in life this way. We are built to make mistakes, coded for error.

We learn, as we say, by "trial and error." Why do we always say that? Why not "trial and rightness" or "trial and triumph"? The old phrase puts it that way because that is, in real life, the way it is done.

A good laboratory, like a good bank or a corporation or government, has to run like a computer. Almost everything is done flawlessly, by the book, and all the numbers add up to the predicted sums. The days go by. And then, if it is a lucky day, and a lucky laboratory, somebody makes a mistake: the wrong buffer, something in one of the blanks, a decimal misplaced in reading counts, the warm room off by a degree and a half, a mouse out of his box, or just a misreading of the day's protocol. Whatever, when the results come in, something is obviously screwed up, and then the action can begin.

The misreading is not the important error; it opens the way. The next step is the crucial one. If the investigator can bring himself to say, "But even so, look at that!" then the new finding, whatever it is, is ready for snatching. What is needed, for progress to be made, is the move based on the error.

Whenever new kinds of thinking are about to be accomplished, or new varieties of music, there has to be an argument beforehand. With two sides debating in the same mind, haranguing, there is an amiable understanding that one is right and the other wrong. Sooner or later the thing is settled, but there can be no action at all if there are not the two sides, and the argument. The hope is in the faculty of wrongness, the tendency toward error. The capacity to leap across mountains of information to land lightly on the wrong side represents the highest of human endowments.

It may be that this is a uniquely human gift, perhaps even stipulated in our genetic instructions. Other creatures do not seem to have DNA sequences for making mistakes as a routine part of daily living, certainly not for programmed error as a guide for action.

We are at our human finest, dancing with our minds, when there are more choices than two. Sometimes there are ten, even twenty different ways to go, all but one bound to be wrong, and the richness of selection in such situations can lift us onto totally new ground. This process is called exploration and is based on human fallibility. If we had only a single center in our brains, capable of responding only when a correct decision was to be made, instead of the jumble of different, credulous, easily conned clusters of neurones that provide for being flung off into blind alleys, up trees, down dead ends, out into blue sky, along wrong turnings, around bends, we could only stay the way we are today, stuck fast.

The lower animals do not have this splendid freedom. They are limited, most of them, to absolute infallibility. Cats, for all their good side, never make mistakes. I have never seen a maladroit, clumsy, or blundering cat. Dogs are sometimes fallible, occasionally able to make charming minor mistakes, but they get this way by trying to mimic their masters. Fish are flawless in everything they do. Individual cells in a tissue are mindless machines,

39

perfect in their performance, as absolutely inhuman as
bees.

We should have this in mind as we become dependent on
more complex computers for the arrangement of our
affairs. Give the computers their heads, I say; let them go
their way. If we can learn to do this, turning our heads to
one side and wincing while the work proceeds, the possibilities for the future of mankind, and computerkind, are
limitless. Your average good computer can make calculations in an instant which would take a lifetime of slide rules
for any of us. Think of what we could gain from the near
infinity of precise, machine-made miscomputation which is
now so easily within our grasp. We could begin the solving
of some of our hardest problems. How, for instance, should
we go about organizing ourselves for social living on a
planetary scale, now that we have become, as a plain fact of
life, a single community? We can assume, as a working
hypothesis, that all the right ways of doing this are unworkable. What we need, then, for moving ahead, is a set of
wrong alternatives much longer and more interesting than
the short list of mistaken courses that any of us can think up
right now. We need, in fact, an infinite list, and when it is
printed out we need the computer to turn on itself and
select, at random, the next way to go. If it is a big enough
mistake, we could find ourselves on a new level, stunned,
out in the clear, ready to move again.

The Selves

THERE ARE psychiatric patients who are said to be incapacitated by having more than one self. One of these, an attractive, intelligent young woman in distress, turned up on a television talk show a while back, sponsored to reveal her selves and their disputes. She possessed, she said, or was possessed by, no fewer than eight other separate women, all different, with different names, arguing and elbowing their way into control of the enterprise, causing unending confusion and embarrassment. She (they) wished to be rid of all of them (her), except of course herself (themselves).

People like this are called hysterics by the professionals, or maybe schizophrenics, and there is, I am told, nothing much that can be done. Having more than one self is supposed to be deeply pathological in itself, and there is no known way to evict trespassers.

I am not sure that the number of different selves is in itself all that pathological; I hope not. Eight strikes me personally as a reasonably small and easily manageable number. It is the simultaneity of their appearance that is the real problem, and I should think psychiatry would do better by simply persuading them to queue up and wait their turn, as happens in the normal rest of us. Couldn't they be conditioned some way, by offering rewards or holding out gently threatening sanctions? "How *do* you do, I'm absolutely delighted to see you here and I have exactly fifty-five minutes, after which I very much regret to say someone else will be dropping in, but could I see you again tomorrow at this same time, do have a chocolate mint and let's just talk, just the two of us." That sort of thing might help at least to get them lined up in some kind of order.

Actually, it would embarrass me to be told that more than a single self is a kind of disease. I've had, in my time, more than I could possibly count or keep track of. The great difference, which keeps me feeling normal, is that mine (ours) have turned up one after the other, on an orderly schedule. Five years ago I was another person, juvenile, doing and saying things I couldn't possibly agree with now. Ten years ago I was a stranger. Twenty–forty years ago . . . I've forgotten. The only thing close to what you might call illness, in my experience, was in the gaps in the queue when one had finished and left the place before the next one was ready to start, and there was nobody around at all. Luckily, that has happened only three or four times that I can recall, once when I'd become a very old child and my adolescent hadn't appeared, and a couple of times later on when there seemed to be some confusion about who was next up. The rest of the time they have waited turns and emerged on cue ready to take over, sometimes breathless and needing last-minute briefing but nonetheless steady enough to go on. The surprising thing has always been how little background information they seemed to need, considering how the times changed. I cannot remember who it was five years ago. He was reading linguistics and had just discovered philology, as I recall, but he left before getting anything much done.

To be truthful there have been a few times when they were all there at once, like those girls on television, clamoring for attention, whole committees of them, a House Committee, a Budget Committee, a Grievance Committee, even a Committee on Membership, although I don't know how any of them ever got in. No chairman, ever, certainly not me. At the most I'm a sort of administrative assistant. There's never an agenda. At the end I bring in the refreshments.

What do we meet about? It is hard to say. The door bangs open and in they come, calling for the meeting to start, and then they all talk at once. Odd to say, it is not just a jumble of talk; they tend to space what they're saying so that words and phrases from one will fit into short spaces left in silence by the others. At good times it has the feel of an intensely complicated conversation, but at others the sounds are more like something overheard in a crowded station. At worse times the silences get out of synchrony, interrupting each other; it is as though all the papers had suddenly blown off the table.

We never get anything settled. In recent years I've sensed an increase in their impatience with me, whoever they think I am, and with the fix they're in. They don't come right out and say so, but what they are beginning to want more than anything else is a chairman.

The worst times of all have been when I've wanted to be just one. Try walking out on the ocean beach at night, looking at stars, thinking, Be one, be one. Doesn't work, ever. Just when you feel ascension, turning, wheeling, and that whirring sound like a mantel clock getting ready to strike, the other selves begin talking. Whatever you're thinking, they say, it's not like that at all.

The only way to quiet them down, get them to stop, is to play music. That does it. Bach stops them every time, in their tracks, almost as though that's what they've been waiting for.

The Health-Care System

THE HEALTH-CARE SYSTEM of this country is a staggering enterprise, in any sense of the adjective. Whatever the failures of distribution and lack of coordination, it is the gigantic scale and scope of the total collective effort that first catches the breath, and its cost. The dollar figures are almost beyond grasping. They vary from year to year, always upward, ranging from something like $10 billion in 1950 to an estimated $140 billion in 1978, with much more to come in the years just ahead, whenever a national health-insurance program is installed. The official guess is that we are now investing a round 8 percent of the GNP in Health; it could soon rise to 10 or 12 percent.

Those are the official numbers, and only for the dollars

that flow in an authorized way—for hospital charges, physician's fees, prescribed drugs, insurance premiums, the construction of facilities, research, and the like.

But these dollars are only part of it. Why limit the estimates to the strictly professional costs? There is another huge marketplace, in which vast sums are exchanged for items designed for the improvement of Health.

The television and radio industry, no small part of the national economy, feeds on Health, or, more precisely, on disease, for a large part of its sustenance. Not just the primarily medical dramas and the illness or surgical episodes threaded through many of the nonmedical stories, in which the central human dilemma is illness; almost all the commercial announcements, in an average evening, are pitches for items to restore failed health: things for stomach gas, constipation, headaches, nervousness, sleeplessness or sleepiness, arthritis, anemia, disquiet, and the despair of malodorousness, sweat, yellowed teeth, dandruff, furuncles, piles. The food industry plays the role of surrogate physician, advertising breakfast cereals as though they were tonics, vitamins, restoratives; they are now out-hawked by the specialized Health-food industry itself, with its nonpolluted, organic, "naturally" vitalizing products. Chewing gum is sold as a tooth cleanser. Vitamins have taken the place of prayer.

The publishing industry, hardcover, paperbacks, magazines, and all, seems to be kept alive by Health, new techniques for achieving mental health, cures for arthritis, and diets mostly for the improvement of everything.

The transformation of our environment has itself become

an immense industry, costing rather more than the moon, in aid of Health. Pollution is supposed to be primarily a medical problem; when the television weatherman tells whether New York's air is "acceptable" or not that day, he is talking about human lungs, he believes. Pollutants which may be impairing photosynthesis by algae in the world's oceans, or destroying all the life in topsoil, or killing all the birds are being worried about lest they cause cancer in us, for heaven's sake.

Tennis has become more than the national sport; it is a rigorous discipline, a form of collective physiotherapy. Jogging is done by swarms of people, out onto the streets each day in underpants, moving in a stolid sort of rapid trudge, hoping by this to stay alive. Bicycles are cures. Meditation may be good for the soul but it is even better for the blood pressure.

As a people, we have become obsessed with Health.

There is something fundamentally, radically unhealthy about all this. We do not seem to be seeking more exuberance in living as much as staving off failure, putting off dying. We have lost all confidence in the human body.

The new consensus is that we are badly designed, intrinsically fallible, vulnerable to a host of hostile influences inside and around us, and only precariously alive. We live in danger of falling apart at any moment, and are therefore always in need of surveillance and propping up. Without the professional attention of a health-care system, we would fall in our tracks.

This is a new way of looking at things, and perhaps it can only be accounted for as a manifestation of spontaneous, undirected, societal *propaganda*. We keep telling each other this sort of thing, and back it comes on television or in the weekly newsmagazines, confirming all the fears, instructing us, as in the usual final paragraph of the personal-advice columns in the daily paper, to "seek professional help." Get a checkup. Go on a diet. Meditate. Jog. Have some surgery. Take two tablets, with water. *Spring* water. If pain persists, if anomie persists, if boredom persists, see your doctor.

It is extraordinary that we have just now become convinced of our bad health, our constant jeopardy of disease and death, at the very time when the facts should be telling us the opposite. In a more rational world, you'd think we would be staging bicentennial ceremonies for the celebration of our general good shape. In the year 1976, out of a population of around 220 million, only 1.9 million died, or just under 1 percent, not at all a discouraging record once you accept the fact of mortality itself. The life expectancy for the whole population rose to seventy-two years, the longest stretch ever achieved in this country. Despite the persisting roster of still-unsolved major diseases—cancer, heart disease, stroke, arthritis, and the rest—most of us have a clear, unimpeded run at a longer and healthier lifetime than could have been foreseen by any earlier generation. The illnesses that plague us the most, when you count up the numbers in the U.S. Vital Statistics reports, are respiratory and gastrointestinal infections, which are, by and large, transient, reversible affairs needing not much more than Grandmother's advice for getting through safely. Thanks in great part to the improved sanitary engineering, nutrition, and housing of the past century, and in real but less part to contemporary immunization and antibiotics, we are free of the great infectious diseases, especially tuberculosis and lobar pneumonia, which used to cut us down long before our time. We are even beginning to make progress in our understanding of the mechanisms underlying the chronic illnesses still with us, and sooner or later, depending on the quality and energy of biomedical research, we will learn to cope effectively with most of these, maybe all. We will still age away and die, but the aging, and even the dying, can become a healthy process. On balance, we ought to be more pleased with ourselves than we are, and more optimistic for the future.

The trouble is, we are being taken in by the propaganda, and it is bad not only for the spirit of society; it will make any health-care system, no matter how large and efficient, unworkable. If people are educated to believe that they are fundamentally fragile, always on the verge of mortal disease, perpetually in need of support by health-care professionals at every side, always dependent on an imagined discipline of "preventive" medicine, there can be no limit to the numbers of doctors' offices, clinics, and hospitals required to meet the demand. In the end, we would all become doctors, spending our days screening each other for disease.

We are, in real life, a reasonably healthy people. Far from being ineptly put together, we are amazingly tough, durable organisms, full of health, ready for most contingencies. The new danger to our well-being, if we continue to listen to all the talk, is in becoming a nation of healthy hypochondriacs, living gingerly, worrying ourselves half to death.

And we do not have time for this sort of thing anymore, nor can we afford such a distraction from our other, considerably more urgent problems. Indeed, we should be worrying that our preoccupation with personal health may be a symptom of copping out, an excuse for running upstairs to recline on a couch, sniffing the air for contaminants, spraying the room with deodorants, while just outside, the whole of society is coming undone.

On Cloning
a Human Being

IT IS NOW theoretically possible to recreate an identical
creature from any animal or plant, from the DNA contained in the nucleus of any somatic cell. A single plant
root-tip cell can be teased and seduced into conceiving a
perfect copy of the whole plant; a frog's intestinal epithelial
cell possesses the complete instructions needed for a new,
same frog. If the technology were further advanced, you
could do this with a human being, and there are now startled predictions all over the place that this will in fact be
done, someday, in order to provide a version of immortality
for carefully selected, especially valuable people.

The cloning of humans is on most of the lists of things to
worry about from Science, along with behavior control,

genetic engineering, transplanted heads, computer poetry, and the unrestrained growth of plastic flowers.

Cloning is the most dismaying of prospects, mandating as it does the elimination of sex with only a metaphoric elimination of death as compensation. It is almost no comfort to know that one's cloned, identical surrogate lives on, especially when the living will very likely involve edging one's real, now aging self off to the side, sooner or later. It is hard to imagine anything like filial affection or respect for a single, unmated nucleus; harder still to think of one's new, self-generated self as anything but an absolute, desolate orphan. Not to mention the complex interpersonal relationship involved in raising one's self from infancy, teaching the language, enforcing discipline, instilling good manners, and the like. How would you feel if you became an incorrigible juvenile delinquent by proxy, at the age of fifty-five?

The public questions are obvious. Who is to be selected, and on what qualifications? How to handle the risks of misused technology, such as self-determined cloning by the rich and powerful but socially objectionable, or the cloning by governments of dumb, docile masses for the world's work? What will be the effect on all the uncloned rest of us of human sameness? After all, we've accustomed ourselves through hundreds of millennia to the continual exhilaration of uniqueness; each of us is totally different, in a fundamental sense, from all the other four billion. Selfness is an essential fact of life. The thought of human nonselfness, precise sameness, is terrifying, when you think about it.

Well, don't think about it, because it isn't a probable possibility, not even as a long shot for the distant future, in my opinion. I agree that you might clone some people who would look amazingly like their parental cell donors, but the odds are that they'd be almost as different as you or me, and certainly more different than any of today's identical twins.

The time required for the experiment is only one of the problems, but a formidable one. Suppose you wanted to clone a prominent, spectacularly successful diplomat, to look after the Middle East problems of the distant future. You'd have to catch him and persuade him, probably not very hard to do, and extirpate a cell. But then you'd have to wait for him to grow up through embryonic life and then for at least forty years more, and you'd have to be sure all observers remained patient and unmeddlesome through his unpromising, ambiguous childhood and adolescence.

Moreover, you'd have to be sure of recreating his environment, perhaps down to the last detail. "Environment" is a word which really means people, so you'd have to do a lot more cloning than just the diplomat himself.

This is a very important part of the cloning problem, largely overlooked in our excitement about the cloned individual himself. You don't have to agree all the way with B. F. Skinner to acknowledge that the environment does make a difference, and when you examine what we really mean by the word "environment" it comes down to other human beings. We use euphemisms and jargon for this, like "social forces," "cultural influences," even Skinner's "verbal community," but what is meant is the dense crowd of nearby people who talk to, listen to, smile or frown at, give to, withhold from, nudge, push, caress, or flail out at the individual. No matter what the genome says, these people have a lot to do with shaping a character. Indeed, if all you had was the genome, and no people around, you'd grow a sort of vertebrate plant, nothing more.

So, to start with, you will undoubtedly need to clone the parents. No question about this. This means the diplomat is out, even in theory, since you couldn't have gotten cells from both his parents at the time when he was himself just recognizable as an early social treasure. You'd have to limit the list of clones to people already certified as sufficiently valuable for the effort, with both parents still alive. The parents would need cloning and, for consistency, their parents as well. I suppose you'd also need the usual informed-consent forms, filled out and signed, not easy to get if I know parents, even harder for grandparents.

But this is only the beginning. It is the whole family that really influences the way a person turns out, not just the parents, according to current psychiatric thinking. Clone the family.

Then what? The way each member of the family develops has already been determined by the environment set around him, and this environment is more people, people outside the family, schoolmates, acquaintances, lovers, enemies, car-pool partners, even, in special circumstances, peculiar strangers across the aisle on the subway. Find them, and clone them.

But there is no end to the protocol. Each of the outer contacts has his own surrounding family, and his and their outer contacts. Clone them all.

To do the thing properly, with any hope of ending up

with a genuine duplicate of a single person, you really have no choice. You must clone the world, no less.

We are not ready for an experiment of this size, nor, I should think, are we willing. For one thing, it would mean replacing today's world by an entirely identical world to follow immediately, and this means no new, natural, spontaneous, random, chancy children. No children at all, except for the manufactured doubles of those now on the scene. Plus all those identical adults, including all of today's politicians, all seen double. It is too much to contemplate.

Moreover, when the whole experiment is finally finished, fifty years or so from now, how could you get a responsible scientific reading on the outcome? Somewhere in there would be the original clonee, probably lost and overlooked, now well into middle age, but everyone around him would be precise duplicates of today's everyone. It would be today's same world, filled to overflowing with duplicates of today's people and their same, duplicated problems, probably all resentful at having had to go through our whole thing all over, sore enough at the clonee to make endless trouble for him, if they found him.

And obviously, if the whole thing were done precisely right, they would still be casting about for ways to solve the problem of universal dissatisfaction, and sooner or later they'd surely begin to look around at each other, wondering who should be cloned for his special value to society, to get us out of all this. And so it would go, in regular cycles, perhaps forever.

I once lived through a period when I wondered what Hell could be like, and I stretched my imagination to try to think of a perpetual sort of damnation. I have to confess, I never thought of anything like this.

I have an alternative suggestion, if you're looking for a way out. Set cloning aside, and don't try it. Instead, go in the other direction. Look for ways to get mutations more quickly, new variety, different songs. Fiddle around, if you must fiddle, but never with ways to keep things the same, no matter who, not even yourself. Heaven, somewhere ahead, has got to be a change.

On Etymons and Hybrids

AN ETYMON is supposed to be a pure ore of a word, crystal-
line, absolutely original, signifying just what it was always
intended to signify. They are very rare these days. Most of
the words we use are hybrids, pieced together out of old,
used speech by a process rather like the recycling of waste.
We keep stores of discarded words around, out beyond the
suburbs of our minds, stacked like scrap metal.

When you do run across a primary, original word, the
experience is both disturbing and vaguely pleasurable, like
coming across a friend's picture in an old high-school annual. They are all very old, and the most meaningful ones
date all the way back to Indo-European roots which became
the parents of cognate words in Sanskrit, Persian, Greek,

Latin, and, much later, most of the English language. *Sen* meant old, *spreg* meant speak, *swem* was swim, *nomen* was name, a *porko* was a young pig, *dent* was a tooth. *Eg* was I and my ego, *tu* was thou, *yu* were you, and *me* was me. *Nek* was death. A *mormor* was a murmur. *Mater, pater, bhrater,* and *swesor* were the immediate family, and *nepots* were the nephews and nieces. A *yero* was a year. A *wopsa* was a wasp and an *apsa* was an aspen. A *deru* was a tree, and also something durable and true. To *gno* was to know. *Akwa* was water, and to *bhreu* was to boil. Using basic Indo-European and waving your hands, you could get around the world almost as well as with New York English.

Some of the first words have changed their meaning drastically, of course. *Bhedh* was the source of our "bead," but it originally meant to ask or bid; "bead" started out as a word for prayer. *Dheye* meant to look and see, and moved from *dhyana* in Sanskrit, meaning to meditate, to *jhana* in Pali, to *ch'an* in Chinese, to *zen* in Japanese.

You'd think modern science would be inventing lots of brand-new etymons to meet its needs, but it is not so. Most of our terms for new things are reconditioned words. "Thermodynamics," first spoken a century ago, is an antique shop: the Indo-European *gwher,* meaning warm, was turned into *thermos* in Greek, while *deu,* to do, became *dunesthai* in Greek, meaning capable of, and hence dynamic (the same *deu* is the source of "dynamite," "bonus," and "bonbon"). A "bit," in computer jargon, although designed as the least ambiguous of terms by combining parts of "binary" and "digit," has a tangle of meanings in its origin: "binary" came from *dwo,* meaning two, which led also to "twig," "double," and "doubt"; "digit" began as *deik,* to show or teach, and moved into English in company with "token," "paradigm," and "ditto," also "toe."

A nucleic acid (from *ken,* later *knu,* plus *ak*) is a sort of nut coupled to something sharp.

"Cholera toxin" could be translated by an outsider new to our language as a bright and shiny bow and arrow. *Ghel* was at first the word for shining, later yellow; it turned to *ghola,* then *khole* in Greek, meaning bile, then into "choler" and "cholera" in English. "Toxin" was originally *tekw,* a word meaning to run or flee, later becoming *toxsa* in Persian and *toxon* in Greek, meaning bow and arrow; the toxin meaning may have come from the poison used to tip the arrows, or, as Robert Graves suggested, from the yew tree *taxus,* from which arrows were best made and whose berries were long thought to be poisonous.

The word for poison came by a devious route, like a long-delayed afterthought. It derives from *poi,* to drink, becoming *potare* in Latin, whence "potion" (and also "symposium," from *sun,* together, plus *posis,* to drink). The venomous meaning did not come until the notion of love potions evolved, and then the idea of poison came to consciousness.

There is the same strange history behind the word "venom." This began as the simple word *wen,* meaning to wish or will, leading more or less directly to "win." Along the way, a fork led to "venus," "venery," and "venerate," all indicating varieties of love. The love potion was called *venin,* and somehow this gradually acquired today's sense of venom.

Nobody can explain why "poison" and "venom" come from love potions. Perhaps it was because the pharmacology of the day was primitive and chancy, a very thin line away from toxicology. Or maybe there was a commonsense consensus that any sort of chemical additive intended to induce false love is, by its nature, a fundamental poison. It tells something important about the good taste of earlier human beings that venom and poison were taken resentfully out of the hands of artificial lovers and transferred to the stings of insects and the fangs of serpents.

A "virus" is a very ancient word, despite the newness of its meaning for us. The root was *weis,* meaning to flow, in the sense of oozing, and it went first from *wase* in Old English to *wose* in Middle English and thence to "ooze" itself. The meaning of something twisty and slippery derived, and the weasel was named. The associations became more unpleasant, indicating something noxious, and "virulence" of "viruses" were the resulting words. "Noxious," incidentally, came from *nek,* meaning death, by way of *necare* and *nocere* in Latin, providing "necropsy" and cognate words for us; nectar was the drink of the gods because it prevented death (*tar,* meaning to overcome).

All of this has the sound of a series of accidents. It may be that the evolution of language was largely a matter of luck, like the evolution of creatures. Even though the facts of the matter have been firmly nailed down by two centuries of meticulous philological scholarship, there is a general, unavoidable sense of high improbability in the whole business. If this is the way words evolved, it seems to have depended upon a lot of pure chance, or, as the French say, hazard.

Chance. Now, there's a word. Partridge gives it almost two columns of the finest print, but not under itself. If you want to look up "chance" you must find "cadence," the nearest thing to an etymon but a long way off at that. "Cadence" comes from *kad,* meaning to fall. *Kad* led to *cadere* in Latin and *cad* in Sanskrit, also meaning to fall, sometimes to die, and from this came a cascade of words with the sense of risk and transiency: "cadaver," "decay," "casualty," "deciduous," and "casuistry."

The notion of falling gave rise to words like "cadence" and "cadenza" and "cascade." The idea of chance came, as you may have guessed, from the falling of dice.

Incidentally, "hazard" also came from dice, by way of Old French *hasard* and Spanish *azar,* from the Arabic *yasara,* to play at dice. The game of dice is named for the die used for playing, and this comes from Indo-European *do,* which originally meant to give, later changing to "donation," "dowry," "endow," "dose," and "antidote." In Vulgar Latin the verb *dare* came to mean play, leading to *datum* as a playing piece, then *dee* in Old English, then "die" and "dice."

It is obvious that this sort of thing could not have been worked out by any intentional human intelligence. Today's language is the result of an interminable series of small blunders, one after another, leading us back through a near infinity of time. The words are simply let loose by all of us, allowed to fly around out there in the dark, bumping into each other, mating in crazy ways, producing wild, random hybrids, beyond the control of reason.

Just think how much better we could manage if we put our minds to it. All it needs is better, clearheaded organization, with a more efficient administrative control of human speech. Management is what has been lacking. If, as seems to be the case, sometimes deplorably, today's most necessary words have been created by this improbable process of hydridization, then hybridization is the business we should now begin to take in hand. All we have to learn is how to pair one word with another so that mating can occur, and then sort out the progeny to our liking. Governments will need to become involved in this, for we shall be requiring whole new institutions all around the earth, occupying huge tracts of land in national capitals, devoted to the breeding of words, like the Agricultural Experimental Stations of the past century. The breeding of words can become the bureaucratic preoccupation for the future, as in the past, but better organized, with more committees. Given a stockpile of innovative in-house creativity for the generation of novel words, substituting numbers for the input of letters wherever feasible, and fiscally optimized by computer capacitization for targeting in on core issues relating to aims, goals and priorities, and learned skills, we might at last be freed from our dependence on the past. New hybrids, synthesized in agencies of our local institutions, could then take the place of those Indo-European words, with all their primitive, precivilized, embarrassing resonances.

To start with, we ought to get another word to take the place of "hybrid." Not that it doesn't describe itself satisfactorily, but there is something not sufficiently straightforward about it for the scientific needs it is intended to serve. "Hybrid" is itself a relatively new word, easily disposed of without sentiment, but standing blank-faced behind it is the Latin *hybrida,* which was the name for the unsuitable off-spring of a wild boar and a domestic sow. The word had no use in English until around the seventeenth century, when a casual mention of hybrids was made, referring to the boar-sow mismatch. It was not until the mid-nineteenth century that it really entered the language. At that time it was needed for botany, zoology, and the rapidly developing discipline of philology, and there were even usages in politics (hybrid bills in Parliament).

The real trouble with "hybrid" is in its more distant origins. It is a word that carries its own disapproval inside. Before being "hybrid" it was *hubris,* an earlier Greek word indicating arrogance, insolence against the gods. *Hubris* itself came from two Indo-European roots, *ud,* meaning up or out, and *gwer,* with the meaning of violence and strength. Outrage was the general sense. *Hubris* became a naturalized English word in the late nineteenth century, exhumed by classical scholars at Oxford and Cambridge, and promptly employed as slang to describe the deliberate use of high intellectual capacity to get oneself into trouble. Hubris was the risk of losing in an academic equivalent of jujitsu; if you used your full mental powers you could be hurled, by your own efforts, out into limbo.

The latest hybrids to join the products of botanists and zoologists are the combinations between the nucleic acids of mammalian and bacterial cells which can be brought about, easy as stringing beads, by the new recombinant-

DNA technology. There are people who wish to stop the manufacture of these hybrids, on grounds that the biological properties of such new beings might be harmful.

Make up our own language? With committees in institutes? What a way to talk.

The Hazards of Science

THE CODE WORD for criticism of science and scientists these days is "hubris." Once you've said that word, you've said it all; it sums up, in a word, all of today's apprehensions and misgivings in the public mind—not just about what is perceived as the insufferable attitude of the scientists themselves but, enclosed in the same word, what science and technology are perceived to be doing to make this century, this near to its ending, turn out so wrong.

"Hubris" is a powerful word, containing layers of powerful meaning, derived from a very old word, but with a new life of its own, growing way beyond the limits of its original meaning. Today, it is strong enough to carry the full weight of disapproval for the cast of mind that thought up atomic fusion and fission as ways of first blowing up and later heating cities as well as the attitudes which led to strip-mining, offshore oil wells, Kepone, food additives, SSTs, and the tiny spherical particles of plastic recently discovered clogging the waters of the Sargasso Sea.

The biomedical sciences are now caught up with physical science and technology in the same kind of critical judgment, with the same pejorative word. Hubris is responsible, it is said, for the whole biological revolution. It is hubris that has given us the prospects of behavior control, psychosurgery, fetal research, heart transplants, the cloning of prominent politicians from bits of their own eminent tissue, iatrogenic disease, overpopulation, and recombinant DNA. This last, the new technology that permits the stitching of one creature's genes into the DNA of another, to make hybrids, is currently cited as the ultimate example of hubris. It is hubris for man to manufacture a hybrid on his own.

So now we are back to the first word again, from "hybrid" to "hubris," and the hidden meaning of two beings joined unnaturally together by man is somehow retained. Today's joining is straight out of Greek mythology: it is the combining of man's capacity with the special prerogative of the gods, and it is really in this sense of outrage that the word "hubris" is being used today. This is what the word has grown into, a warning, a code word, a shorthand signal from the language itself: if man starts doing things reserved for the gods, deifying himself, the outcome will be something worse for him, symbolically, than the litters of wild boars and domestic sows were for the ancient Romans.

To be charged with hubris is therefore an extremely serious matter, and not to be dealt with by murmuring things about antiscience and antiintellectualism, which is what many of us engaged in science tend to do these days. The doubts about our enterprise have their origin in the most profound kind of human anxiety. If we are right and the critics are wrong, then it has to be that the word "hubris" is being mistakenly employed, that this is not what we are up to, that there is, for the time being anyway, a fundamental misunderstanding of science.

I suppose there is one central question to be dealt with, and I am not at all sure how to deal with it, although I am quite certain about my own answer to it. It is this: are there some kinds of information leading to some sorts of knowledge that human beings are really better off not having? Is there a limit to scientific inquiry not set by what is knowable but by what we *ought* to be knowing? Should we stop short of learning about some things, for fear of what we, or someone, will do with the knowledge? My own answer is a flat no, but I must confess that this is an intuitive response and I am neither inclined nor trained to reason my way through it.

There has been some effort, in and out of scientific quarters, to make recombinant DNA into the issue on which to settle this argument. Proponents of this line of research are accused of pure hubris, of assuming the rights of gods, of arrogance and outrage; what is more, they confess themselves to be in the business of making live hybrids with their own hands. The mayor of Cambridge and the attorney general of New York have both been advised to put a stop to it, forthwith.

It is not quite the same sort of argument, however, as the one about limiting knowledge, although this is surely part of it. The knowledge is already here, and the rage of the argument is about its application in technology. Should DNA for making certain useful or interesting proteins be incorporated into *E. coli* plasmids or not? Is there a risk of inserting the wrong sort of toxins or hazardous viruses, and then having the new hybrid organisms spread beyond the laboratory? Is this a technology for creating new varieties of pathogens, and should it be stopped because of this?

If the argument is held to this level, I can see no reason why it cannot be settled, by reasonable people. We have learned a great deal about the handling of dangerous microbes in the last century, although I must say that the opponents of recombinant-DNA research tend to downgrade this huge body of information. At one time or another, agents as hazardous as those of rabies, psittacosis, plague, and typhus have been dealt with by investigators in secure laboratories, with only rare instances of self-infection of the investigators themselves, and no instances at all of epidemics. It takes some high imagining to postulate the creation of brand-new pathogens so wild and voracious as to spread from equally secure laboratories to endanger human life at large, as some of the arguers are now maintaining.

But this is precisely the trouble with the recombinant-DNA problem: it has become an emotional issue, with too many irretrievably lost tempers on both sides. It has lost the sound of a discussion of technological safety, and begins now to sound like something else, almost like a religious controversy, and here it is moving toward the central issue: are there some things in science we should not be learning about?

There is an inevitably long list of hard questions to follow this one, beginning with the one which asks whether the mayor of Cambridge should be the one to decide, first off.

Maybe we'd be wiser, all of us, to back off before the recombinant-DNA issue becomes too large to cope with. If we're going to have a fight about it, let it be confined to the immediate issue of safety and security, of the recombinants now under consideration, and let us by all means have regulations and guidelines to assure the public safety wherever these are indicated or even suggested. But if it is possible let us stay off that question about limiting human knowledge. It is too loaded, and we'll simply not be able to cope with it.

By this time it will have become clear that I have already taken sides in the matter, and my point of view is entirely prejudiced. This is true, but with a qualification. I am not so much in favor of recombinant-DNA research as I am opposed to the opposition to this line of inquiry. As a longtime student of infectious-disease agents I do not take kindly the declarations that we do not know how to keep from catching things in laboratories, much less how to keep them from spreading beyond the laboratory walls. I believe we learned a lot about this sort of thing, long ago. Moreover, I regard it as a form of hubris-in-reverse to claim that man can make deadly pathogenic microorganisms so easily. In my view, it takes a long time and a great deal of interliving before a microbe can become a successful pathogen. Pathogenicity is, in a sense, a highly skilled trade, and only a tiny minority of all the numberless tons of microbes on the earth has ever been involved itself in it; most bacteria are busy with their own business, browsing and recycling the rest of life. Indeed, pathogenicity often seems to me a sort of biological accident in which signals are misdirected by the microbe or misinterpreted by the host, as in the case of endotoxin, or in which the intimacy between host and microbe is of such long standing that a form of molecular mimicry becomes possible, as in the case of diphtheria toxin. I do not believe that by simply putting together new combinations of genes one can create creatures as highly skilled and adapted for dependence as a pathogen must be, any more than I have ever believed that microbial life from the moon or Mars could possibly make a living on this planet.

But, as I said, I'm not at all sure this is what the argument is really about. Behind it is that other discussion, which I wish we would not have to become enmeshed in.

I cannot speak for the physical sciences, which have moved an immense distance in this century by any standard, but it does seem to me that in the biological and medical sciences we are still far too ignorant to begin making judgments about what sorts of things we should be learning or not learning. To the contrary, we ought to be grateful for whatever snatches we can get hold of, and we ought to be out there on a much larger scale than today's, looking for more.

We should be very careful with that word "hubris," and make sure it is not used when not warranted. There is a great danger in applying it to the search for knowledge. The application of knowledge is another matter, and there is hubris in plenty in our technology, but I do not believe that looking for new information about nature, at whatever level, can possibly be called unnatural. Indeed, if there is any single attribute of human beings, apart from language, which distinguishes them from all other creatures on earth, it is their insatiable, uncontrollable drive to learn things and then to exchange the information with others of the species. Learning is what we do, when you think about it. I cannot think of a human impulse more difficult to govern.

But I can imagine lots of reasons for trying to govern it. New information about nature is very likely, at the outset, to be upsetting to someone or other. The recombinant-DNA line of research is already upsetting, not because of the dangers now being argued about but because it is disturbing, in a fundamental way, to face the fact that the genetic machinery in control of the planet's life can be fooled around with so easily. We do not like the idea that anything so fixed and stable as a species line can be changed. The notion that genes can be taken out of one genome and inserted in another is unnerving. Classical mythology is peopled with mixed beings—part man, part animal or plant —and most of them are associated with tragic stories. Recombinant DNA is a reminder of bad dreams.

The easiest decision for society to make in matters of this

kind is to appoint an agency, or a commission, or a subcommittee within an agency to look into the problem and provide advice. And the easiest course for a committee to take, when confronted by any process that appears to be disturbing people or making them uncomfortable, is to recommend that it be stopped, at least for the time being.

I can easily imagine such a committee, composed of unimpeachable public figures, arriving at the decision that the time is not quite ripe for further exploration of the transplantation of genes, that we should put this off for a while, maybe until next century, and get on with other affairs that make us less discomfited. Why not do science on something more popular, say, how to get solar energy more cheaply? Or mental health?

The trouble is, it would be very hard to stop once this line was begun. There are, after all, all sorts of scientific inquiry that are not much liked by one constituency or another, and we might soon find ourselves with crowded rosters, panels, standing committees, set up in Washington for the appraisal, and then the regulation, of research. Not on grounds of the possible value and usefulness of the new knowledge, mind you, but for guarding society against scientific hubris, against the kinds of knowledge we're better off without.

It would be absolutely irresistible as a way of spending time, and people would form long queues for membership. Almost anything would be fair game, certainly anything to do with genetics, anything relating to population control, or, on the other side, research on aging. Very few fields would get by, except perhaps for some, like mental health, in which nobody really expects anything much to happen, surely nothing new or disturbing.

The research areas in the greatest trouble would be those already containing a sense of bewilderment and surprise, with discernible prospects of upheaving present dogmas.

It is hard to predict how science is going to turn out, and if it is really good science it is impossible to predict. This is in the nature of the enterprise. If the things to be found are actually new, they are by definition unknown in advance, and there is no way of telling in advance where a really new line of inquiry will lead. You cannot make choices in this matter, selecting things you think you're going to like and shutting off the lines that make for discomfort. You either have science or you don't, and if you have it you are obliged to accept the surprising and disturbing pieces of information, even the overwhelming and upheaving ones, along with the neat and promptly useful bits. It is like that.

The only solid piece of scientific truth about which I feel totally confident is that we are profoundly ignorant about nature. Indeed, I regard this as the major discovery of the past hundred years of biology. It is, in its way, an illuminating piece of news. It would have amazed the brightest minds of the eighteenth-century Enlightenment to be told by any of us how little we know, and how bewildering seems the way ahead. It is this sudden confrontation with the depth and scope of ignorance that represents the most significant contribution of twentieth-century science to the human intellect. We are, at last, facing up to it. In earlier times, we either pretended to understand how things worked or ignored the problem, or simply made up stories to fill the gaps. Now that we have begun exploring in earnest, doing serious science, we are getting glimpses of how huge the questions are, and how far from being answered. Because of this, these are hard times for the human intellect, and it is no wonder that we are depressed. It is not so bad being ignorant if you are totally ignorant; the hard thing is knowing in some detail the reality of ignorance, the worst spots and here and there the not-so-bad spots, but no true light at the end of any tunnel nor even any tunnels that can yet be trusted. Hard times, indeed.

But we are making a beginning, and there ought to be some satisfaction, even exhilaration, in that. The method works. There are probably no questions we can think up that can't be answered, sooner or later, including even the matter of consciousness. To be sure, there may well be questions we can't think up, ever, and therefore limits to the reach of human intellect which we will never know about, but that is another matter. Within our limits, we should be able to work our way through to all our answers, if we keep at it long enough, and pay attention.

I am putting it this way, with all the presumption and confidence that I can summon, in order to raise another, last question. Is this hubris? Is there something fundamentally unnatural, or intrinsically wrong, or hazardous for the species in the ambition that drives us all to reach a comprehensive understanding of nature, including ourselves? I cannot believe it. It would seem to me a more unnatural thing, and more of an offense against nature, for us to come on the same scene endowed as we are with curiosity, filled to overbrimming as we are with questions, and naturally talented as we are for the asking of clear questions, and then for us to do nothing about it or, worse, to try to suppress the questions. This is the greater danger for our species, to try to pretend that we are another kind of animal, that we do not need to satisfy our curiosity, that we can get along somehow without inquiry and exploration and experimentation, and that the human mind can rise above its ignorance by simply asserting that there are things it has no need to know. This, to my way of thinking, is the real hubris, and it carries danger for us all.

On Warts

WARTS ARE wonderful structures. They can appear overnight on any part of the skin, like mushrooms on a damp lawn, full grown and splendid in the complexity of their architecture. Viewed in stained sections under a microscope, they are the most specialized of cellular arrangements, constructed as though for a purpose. They sit there like turreted mounds of dense, impenetrable horn, impregnable, designed for defense against the world outside.

In a certain sense, warts are both useful and essential, but not for us. As it turns out, the exuberant cells of a wart are the elaborate reproductive apparatus of a virus.

You might have thought from the looks of it that the cells infected by the wart virus were using this response as a ponderous way of defending themselves against the virus, maybe even a way of becoming more distasteful, but it is not so. The wart is what the virus truly wants; it can flourish only in cells undergoing precisely this kind of overgrowth. It is not a defense at all; it is an overwhelming welcome, an enthusiastic accommodation meeting the needs of more and more virus.

The strangest thing about warts is that they tend to go away. Fully grown, nothing in the body has so much the look of toughness and permanence as a wart, and yet, inexplicably and often very abruptly, they come to the end of their lives and vanish without a trace.

And they can be made to go away by something that can only be called thinking, or something like thinking. This is a special property of warts which is absolutely astonishing, more of a surprise than cloning or recombinant DNA or endorphin or acupuncture or anything else currently attracting attention in the press. It is one of the great mystifications of science: warts can be ordered off the skin by hypnotic suggestion.

Not everyone believes this, but the evidence goes back a long way and is persuasive. Generations of internists and dermatologists, and their grandmothers for that matter, have been convinced of the phenomenon. I was once told by a distinguished old professor of medicine, one of Sir William Osler's original bright young men, that it was his practice to paint gentian violet over a wart and then assure the patient firmly that it would be gone in a week, and he never saw it fail. There have been several meticulous studies by good clinical investigators, with proper controls. In one of these, fourteen patients with seemingly intractable generalized warts on both sides of the body were hypnotized, and the suggestion was made that all the warts on one side of the body would begin to go away. Within several weeks the results were indisputably positive; in nine patients, all or nearly all of the warts on the suggested side had vanished, while the control side had just as many as ever.

It is interesting that most of the warts vanished precisely as they were instructed, but it is even more fascinating that mistakes were made. Just as you might expect in other affairs requiring a clear understanding of which is the right and which the left side, one of the subjects got mixed up and destroyed the warts on the wrong side. In a later study by a group at the Massachusetts General Hospital, the warts on both sides were rejected even though the instructions were to pay attention to just one side.

I have been trying to figure out the nature of the instructions issued by the unconscious mind, whatever that is, under hypnosis. It seems to me hardly enough for the mind to say, simply, get off, eliminate yourselves, without providing something in the way of specifications as to how to go about it.

I used to believe, thinking about this experiment when it was just published, that the instructions might be quite simple. Perhaps nothing more detailed than a command to shut down the flow through all the precapillary arterioles in and around the warts to the point of strangulation. Exactly how the mind would accomplish this with precision, cutting off the blood supply to one wart while leaving others intact, I couldn't figure out, but I was satisfied to leave it there anyhow. And I was glad to think that my unconscious mind would have to take the responsibility for this, for if I had been one of the subjects I would never have been able to do it myself.

But now the problem seems much more complicated by the information concerning the viral etiology of warts, and even more so by the currently plausible notion that immunologic mechanisms are very likely implicated in the rejection of warts.

If my unconscious can figure out how to manipulate the mechanisms needed for getting around that virus, and for deploying all the various cells in the correct order for tissue rejection, then all I have to say is that my unconscious is a lot further along than I am. I wish I had a wart right now, just to see if I am that talented.

There ought to be a better word than "Unconscious," even capitalized, for what I have, so to speak, in mind. I was brought up to regard this aspect of thinking as a sort of private sanitarium, walled off somewhere in a suburb of my brain, capable only of producing such garbled information as to keep my mind, my proper Mind, always a little off balance.

But any mental apparatus that can reject a wart is something else again. This is not the sort of confused, disordered process you'd expect at the hands of the kind of Unconscious you read about in books, out at the edge of things making up dreams or getting mixed up on words or having hysterics. Whatever, or whoever, is responsible for this has the accuracy and precision of a surgeon. There almost has to be a Person in charge, running matters of meticulous detail beyond anyone's comprehension, a skilled engineer and manager, a chief executive officer, the head of the whole place. I never thought before that I possessed such a tenant. Or perhaps more accurately, such a landlord, since I would be, if this is in fact the situation, nothing more than a lodger.

Among other accomplishments, he must be a cell biologist of world class, capable of sorting through the various classes of one's lymphocytes, all with quite different functions which I do not understand, in order to mobilize the right ones and exclude the wrong ones for the task of tissue rejection. If it were left to me, and I were somehow empowered to call up lymphocytes and direct them to the vicinity of my wart (assuming that I could learn to do such a thing), mine would come tumbling in all unsorted, B cells and T cells, suppressor cells and killer cells, and no doubt other cells whose names I have not learned, incapable of getting anything useful done.

Even if immunology is not involved, and all that needs doing is to shut off the blood supply locally, I haven't the faintest notion how to set that up. I assume that the selective turning off of arterioles can be done by one or another chemical mediator, and I know the names of some of them, but I wouldn't dare let things like these loose even if I knew how to do it.

Well, then, who does supervise this kind of operation? Someone's got to, you know. You can't sit there under hypnosis, taking suggestions in and having them acted on with such accuracy and precision, without assuming the existence of something very like a controller. It wouldn't do to fob off the whole intricate business on lower centers without sending along a quite detailed set of specifications, way over my head.

Some intelligence or other knows how to get rid of warts, and this is a disquieting thought.

It is also a wonderful problem, in need of solving. Just think what we would know, if we had anything like a clear understanding of what goes on when a wart is hypnotized away. We would know the identity of the cellular and chemical participants in tissue rejection, conceivably with some added information about the ways that viruses create foreignness in cells. We would know how the traffic of these reactants is directed, and perhaps then be able to understand the nature of certain diseases in which the traffic is being conducted in wrong directions, aimed at the wrong cells. Best of all, we would be finding out about a kind of superintelligence that exists in each of us, infinitely smarter and possessed of technical know-how far beyond our present understanding. It would be worth a War on Warts, a Conquest of Warts, a National Institute of Warts and All.

On Transcendental Metaworry (TMW)

IT IS SAID that modern, industrialized, civilized human beings are uniquely nervous and jumpy, unprecedentedly disturbed by the future, despaired by the present, sleepless at memories of the recent past, all because of the technological complexity and noisiness of the machinery by which we are surrounded, and the rigidified apparatus of cold steel and plastic which we have constructed between ourselves and the earth. Incessant worry, according to this view, is a modern invention. To turn it off, all we need do is turn off the engines and climb down into the countryside. Primitive man, rose-garlanded, slept well.

I doubt this. Man has always been a specifically anxious creature with an almost untapped capacity for worry; it is a gift that distinguishes him from other forms of life. There is undoubtedly a neural center deep in the human brain for mediating this function, like the centers for hunger or sleep.

Prehistoric man, without tools or fire to be thinking about, must have been the most anxious of us all. Fumbling about in dimly lit caves, trying to figure out what he ought really to be doing, sensing the awesome responsibilities for toolmaking just ahead, he must have spent a lot of time contemplating his thumbs and fretting about them. I can imagine him staring at his hands, apposing thumbtips to each fingertip in amazement, thinking, By God, that's something to set us apart from the animals—and then the grinding thought, What on earth are they for? There must have been many long, sleepless nights, his mind all thumbs.

It would not surprise me to learn that there were ancient prefire committees, convened to argue that thumbs might be taking us too far, that we'd have been better off with simply another finger of the usual sort.

Worrying is the most natural and spontaneous of all human functions. It is time to acknowledge this, perhaps even to learn to do it better. Man is the Worrying Animal. It is a trait needing further development, awaiting perfection. Most of us tend to neglect the activity, living precariously out on the thin edge of anxiety but never plunging in.

For total immersion in the experience of pure, illuminating harassment, I can recommend a modification of the technique of Transcendental Meditation, which I stumbled across after reading an article on the practice in a scholarly magazine and then trying it on myself, sitting on an overturned, stove-in canoe under a beech tree in my backyard.

Following closely the instructions, I relaxed, eyes closed, breathing regularly, repeating a recommended mantra, in this instance the word "oom," over and over. The conditions were suitable for withdrawal and detachment; my consciousness, which normally spends its time clutching for any possible handhold, was prepared to cut adrift. Then, suddenly, the telephone began to ring inside the house, rang several times between breathed "oom"'s, and stopped. In the instant, I discovered Transcendental Worry.

Transcendental Worry can be engaged in at any time, by anyone, regardless of age, sex, or occupation, and in almost any circumstance. For beginners, I advise twenty-minute sessions, in the morning before work and late in the evening just before insomnia.

What you do is sit down someplace, preferably by yourself, and tense all muscles. If you make yourself reasonably uncomfortable at the outset, by sitting on a canoe bottom, say, the tension will come naturally. Now close the eyes, concentrate on this until the effort causes a slight tremor of the eyelids. Now breathe, thinking analytically about the muscular effort involved; it is useful to attempt breathing through one nostril at a time, alternating sides.

Now, the mantra. The word "worry," repeated quite rapidly, is itself effective, because of the allusive cognates in its history. Thus, intruding into the recitation of the mantra comes the recollection that it derives from the Indo-European root *wer,* meaning to turn or bend in the sense of evading, which became *wyrgan* in Old English, meaning to kill by strangling, with close relatives "weird," "writhe," "wriggle," "wrestle," and "wrong." "Wrong" is an equally useful mantra, for symmetrical reasons.

Next, try to float your consciousness free. You will feel something like this happening after about three minutes, and, almost simultaneously with the floating, yawing and sinking will begin. This complex of conjoined sensations becomes an awareness of concentrated, irreversible trouble.

Finally you will begin to hear the *zing,* if you are successful. This is a distant, rhythmic sound, not timed with either the breathing or the mantra. After several minutes, you will discover by taking your pulse that the *zing* is synchronous, and originates somewhere in the lower part of the head or perhaps high up in the neck, presumably due to turbulence at the bend of an artery, maybe even the vibration of a small plaque. Now you are In Touch.

Nothing remains but to allow the intensification of Transcendental Worry to proceed spontaneously to the next stage, termed the Primal Wince. En route, you pass through an almost confluent series of pictures, random and transient, jerky and running at overspeed like an old movie, many of them seemingly trivial but each associated with a sense of dropping abruptly through space (it is useful, here, to recall that "vertigo" also derives from *wer*). You may suddenly see, darting across the mind like a shrieking plumed bird, a current electric-light bill, or the vision of numbers whirring too fast to read on a gasoline pump, or the last surviving humpback whale, singing a final song into empty underseas, or simply the television newscast announcing that détente now signifies a Soviet-American Artificial-Heart Project. Or late bulletins from science concerning the pulsing showers of neutrino particles, aimed personally by collapsing stars, which cannot be escaped from even at the bottom of salt mines in South Dakota. Watergate, of course. The music of John Cage. The ascending slopes of chalked curves on academic blackboards, interchangeably predicting the future population of pet dogs in America, rats in Harlem, nuclear explosions overhead and down in salt mines, suicides in Norway, crop failures in India, the number of people at large. The thought of moon gravity as a cause of baldness. The unpreventability of continental drift. The electronic guitar. The slipping away of things, the feel of rugs sliding out from under everywhere. These images become confluent and then amorphous, melting together into a solid, gelatinous thought of skewness. When this happens, you will be entering the last stage, which is pure worry about pure worry. This is the essence of the Wisdom of the West, and I shall call it Transcendental Metaworry (TMW).

Now, as to the usefulness of TMW. First of all, it tends to fill the mind completely at times when it would otherwise be empty. Instead of worrying at random, continually and subliminally, wondering always what it is that you've forgotten and ought to be worrying about, you get the full experience, all in a rush, on a schedule which you arrange for yourself.

Secondly, it makes the times of the day when there is really nothing to worry about intensely pleasurable, because of the contrast.

Thirdly, I have forgotten the third advantage, which is itself one less thing to worry about.

There are, of course, certain disadvantages, which must be faced up to. TMW is, admittedly, a surrogate experience, a substitute for the real thing, and in this sense there is always the danger of overdoing it. Another obvious danger is the likely entry of technology into the field. I have no doubt that there will soon be advertisements in the back pages of small literary magazines, offering for sale, money back if dissatisfied (or satisfied), electronic devices encased in black plastic boxes with dials, cathode screens, earphones with simulated sonic booms, and terminals to be affixed at various areas of the scalp so that brain waves associated with pure TMW can be identified and volitionally selected. These will be marketed under attractive trade names, like the Angst Amplifier or the Artificial Heartsink. The thought of such things is something else to worry about, but perhaps not much worse than the average car radio.

An Apology

THE ROLE PLAYED by the observer in biological research is complicated but not bizarre: he or she simply observes, describes, interprets, maybe once in a while emits a hoarse shout, but that is that; the act of observing does not alter fundamental aspects of the things observed, or anyway isn't supposed to.

It is very different in modern physics. The uncertainty principle doesn't mean that the observer necessarily destroys the precise momentum, or shifts the particle, in the act of observing, although these things happen. It is a more profound effect. The observer, and his apparatus, *create* the reality to be observed. Without him, there are all sorts of possibilities for single particles, in all sorts of wave patterns.

The reality to be studied by his instruments is not simply there; it is brought into existence by the laboratory.

I got to thinking about this, but couldn't hold my mind still long enough. Words kept getting in the way. The glossary of physics is an enchantment in itself: "charm," "strangeness," "strong" and "weak" forces, "quarks." "Matter" is rather a dreamy word in itself, growing out of an Indo-European root based on baby talk: *ma,* which became *mater,* later differentiating into words like "maternal," "material," and "matrix." Demeter's name came from this root, when she was the goddess of the whole earth.

Then, while I was thinking about this, I suddenly remembered that I've been doing some physical observing on my own, without formal training and with only a pencil point as instrument, and perhaps I've caused trouble without intending to. I did not mean to change things, and I would like to say that I am sorry for the disturbance, if there was a disturbance.

Several times, in the past year or so, I have sat at my desk in an upper-floor room facing north on East Sixty-ninth Street, looking straight at the reflections of the sun in one or another windowpane of a tall apartment building on Seventy-second and Third Avenue. The panes where the sun appears, around early afternoon, vary slowly with the season, as you might expect, and much more quickly with the time of day. If I look long enough, I can carry as many as eight yellow-green afterimages of the sun, place them wherever I want to on my wall, and move them up or down, all eight suns, at will.

Now, I have to say what I've been doing.

On occasion, I place my pencil point (this is best done with a well-sharpened pencil) in the middle of my paper (I write on a yellow lined pad) in the center of my desk, keeping an eye on the Seventy-second and Third apartment, and then I hold the point there.

What I do, at these times, is to change the way the system works. Instead of having the earth rotate around itself every twenty-four hours, I hold the pencil point firmly and make the sun revolve slowly around East Sixty-ninth Street. Anyone can do this. It takes a bit of heaving to get it started, but after a few minutes of hard thought you can hold East Sixty-ninth as the still, central point, and then you can feel the sun rolling up behind you from the right side, making the great circle around. Once you've got the sun started, it is not too difficult to organize the rest of the solar system, so that the whole apparatus is circling around an immobile, still earth and, more specifically, spinning around a central point on the Upper East Side of Manhattan. There are some eccentricities and asymmetries to cope with, to be sure, and it is not a tidy event, but it works nonetheless.

What I did not realize when I began to do this was that it entailed, necessarily, more than just the solar system.

You have to get the galaxy swinging around, all the way round, in just twenty-four hours, and then there are all the other galaxies, which cannot be left dangling out there. They must be swung at the same time, in exact pace with the local sun, and at the same time, while they are being heaved around, whistling through the turbulence of solar wind, they have to be permitted their own frictionless, rhythmic dance around each other, with their own components dancing within their interiors. It is an immense task, and you must hold the pencil point firmly to get it done right. You have to do the whole thing, completely, or you will shake the structure to pieces.

If you want the sun to revolve around the earth in a complete turn every twenty-four hours, you must bring along the whole universe, all the galaxies, all the items in space, clear out to the curved edge.

The hardest part of it is the speed you need to swing around the outermost galaxies, so that everything goes round within the twenty-four hours. What it means is that you need enormous rates of travel, far beyond the speed of light, or you'll have parts dawdling, hanging behind at the outer edges, and it won't work. The universe will swing around the fixed, motionless earth every twenty-four hours, but you must be willing to put the time in on it and keep a firm hold on the pencil.

What bothers me now is the effect this may have had on the cosmologists, who may be looking at things in Pasadena, or Puerto Rico, or Palomar, or Pittsburgh, or wherever. It is probably all right during the times when I have been swinging the universe around, assuming that I'm doing it all of a piece, and that there are, in fact, no membranes of attachment at the edges which I may, unwittingly, be tearing loose. But what happens when I become tired of it, which I do, and let the pencil go, and go off to think about something else? I should think there must be some sort of lurch, some tremor all the way out to the edge, while the readjustment is made to the old way, with the earth turning over and over every twenty-four hours all by itself, and swinging around the sun.

I thought I should say something about this, in case some corrections in the observations are needed for the times when I've done it. But also, it occurs to me that my manipulations may not be the only ones going on. It is entirely possible, now that I think about it, that there is someone over on Central Park West, in his apartment, swinging the universe around a still point in the upper Eighties. Or even someone in Teaneck. Or, skewing everything beyond my comprehension, farther west, even out to San Francisco. This may, in fact, be going on all the time, heaving the universe this way and that, around one still point after another, sometimes even pulling against each other, and the astronomers should certainly be told of this before it is too late to make sense of the mess of numbers.

I am sorry to have done this, myself, but this does not mean that I can be sure of stopping. Once you have held the pencil point with all that precision, on a single fine point, and swung the whole whistling universe around that point, shrinking celestial masses of matter to nothing at all in the necessary speed, feeling the whole thing yaw and heave and almost spin off beyond control, but still holding it there, spinning, it is hard to stop.

On Disease

THE MENINGOCOCCUS, viewed from a distance, seems to have the characteristics of an implacable, dangerous enemy of the whole human race. Epidemics sweep through military barracks, across schoolyards, sometimes over the populations of whole cities. The organism invades the bloodstream, then the meningeal space; the outcome is meningitis, a formidable and highly fatal affliction in the days before chemotherapy. The engagement has the look of specificity, in the sense that the meningococcus appears to be particularly adapted for life in the meninges of human beings. You might even say it makes its living this way, a predator with us as prey.

But it is not so. When you count up the total number of

people infected by the meningococcus, and then compare this with the number coming down with meningitis, the arrangement has a quite different look. The cases of actual meningitis are always a very small minority. There is an infection of the majority, to be sure, but it is confined to the rhinopharynx and usually goes unnoticed by the infected people; they produce antimeningococcal antibodies in their blood a few days after the infection, and the organisms may or may not persist in the pharyngeal mucosa, but that is the end of the affair; there is no invasion of the central nervous system.

The cases of meningitis are the exception. The rule for meningococcal infection is a benign, transient infection of the upper respiratory tract, hardly an infection at all, more like an equable association. It is still a mystery that meningitis develops in some patients, but it is unlikely that this represents a special predilection of the bacteria; it may be that the defense mechanisms of affected patients are flawed in some special way, so that the meningococci are granted access, invited in, so to say. Whatever, the disease is a sort of abnormal event in nature, rather like an accident.

The virus of lymphocytic choriomeningitis is ubiquitous among mice. The classical disease is a lethal form of meningitis, in which the exudate over the surface of the brain is composed entirely of lymphocytes. At first glance, the disease seems to represent invasion of, and damage to, the central nervous system by a virus specifically adapted for such behavior. In actual fact, however, the disease is caused by invasion of the brain surface by the host's own lymphocytes, rather than by any neurotoxic property of the virus. If the lymphocyte response is prevented—as, for example, by inducing infection during fetal life so that "tolerance" to the virus occurs—the outcome is a persistent virus infection everywhere, including the central nervous system, but without any evidence of brain disease. If the immunological response is now restored, by implanting lymphoid tissue from normal, nontolerant mice, meningitis then occurs within a few days. The new lymphocytes swarm over the surface of the brain, looking for the virus, and this is fatal. The disease is, essentially, the result of the host's response to the virus.

Cortisone, which has among its numerous properties the capacity to turn off various defense reactions against bacteria, also seems to turn off the most conspicuous clinical manifestations of infectious disease. Finland, in the early 1950s, shortly after cortisone became available for clinical research, treated several patients with pneumococcal lobar pneumonia and primary atypical pneumonia with cortisone, and observed what seemed at first a miraculous clinical cure. Within a few hours the fever, malaise, prostration, chest pain, and cough vanished, and the patients felt themselves to be restored to abundant good health, asking for dinner and claiming to be able to be up and around. At the same time, however, the X-ray evidence of pneumonia showed an alarming extension of the process, and the experiment was promptly terminated. Subsequently, others observed a similar dramatic elimination of disease manifestations in typhoid fever and rickettsial infection, also associated with the unacceptable trade-off of enhanced spread of infection.

The most spectacular examples of host governance of disease mechanisms are the array of responses elicited in various animals by the lipopolysaccharide endotoxins of gram-negative bacteria. Here the microbial toxin does not even seem to be, in itself, toxic. Although the material has powerful effects on various cells and tissues, including polymorphonuclear leucocytes, platelets, lymphocytes, macrophages, arteriolar smooth muscle, and on complement and the coagulation mechanism, all of these effects represent perfectly normal responses, things done every day in the normal course of living. What makes it a disaster is that they are all turned on at once by the host, as though in response to an alarm signal, and the outcome is widespread tissue destruction, as in the generalized Shwartzman reaction, or outright failure of the circulation of blood, as in endotoxin shock.

The Shwartzman reaction can be prevented by simply lifting out one of the participants in the response. This is done by removing the polymorphonuclear leucocytes temporarily, by treatment with nitrogen mustard, or by preventing blood coagulation with heparin. Animals so treated are unable to develop either the local or generalized Shwartzman reaction. The phenomena of lethal shock can be totally prevented by prior treatment with cortisone.

It is not known how endotoxin acts to produce its signal, but the mechanism seems to be a very old one in nature. One of the most sensitive of all experimental animals is the horseshoe crab, *Limulus polyphemus,* in which an injection into the bloodstream of 1 microgram of lipopolysaccharide will cause a violent response. The circulating hemocytes become enmeshed in dense aggregates, bound up in a coagulated protein which is secreted by these cells, with the result that the flow of blood comes to a standstill and the animal dies. What this seems to represent is an enormously exaggerated defense reaction, aimed at protecting *Limulus* against invasion by gram-negative pathogens. Frederick Bang has demonstrated that the hemocyte granules contain a coagulable protein, which is extruded when gram-negative bacteria enter the tissues; normally, one assumes, individual microorganisms are entrapped in this way and subsequently phagocytosed. When the purified endotoxin is injected into the blood, this becomes propaganda, information that bacteria are everywhere, needing entrapment, and all the hemocytes extrude the protein forthwith. Indeed, there is now evidence that the signal of endotoxin is received, directly, by a receptor contained in extracts of the hemocytes; hence the exquisitely sensitive method for assaying endotoxin in the presence of *Limulus* hemocyte extract, in which coagulation is produced by lipopolysaccharide in concentrations as low as 1 nanogram per milliliter.

From the horseshoe crab's point of view, this is no doubt a valuable and efficient mechanism for the purpose of keeping out pathogens. When it works well, against single microorganisms or small clusters, it carries no hazard. But when barriers are breached and bacteria appear in large numbers, or when purified endotoxin is injected in the laboratory, it becomes an expensive kind of defense. Thus, the defense mechanism becomes itself the disease and the cause of death, while the bacteria play the role of bystanders, innocent from their viewpoint.

Even when bacteria are frontally toxic and destructive for the cells of the host, as in the case of organisms which elaborate exotoxins, there is some question as to the directness of the encounter. The diphtheria bacillus would not be in any sense a pathogen were it not for its toxin, but the toxin-cell reaction must be a two-way relationship of great intimacy, involving the recognition and fitting precisely into the molecular machinery of the cell, as though the toxin were being mistaken for a normal participant in protein synthesis. Moreover, the toxin is not, properly speaking, the diphtheria bacillus' own idea; it is made by the bacterium under instructions from a virus, the bacteriophage. Only the organisms that have become lysogenic for the virus are toxigenic. Dipththeria is not simply an infection by the diphtheria bacillus; it is an infection by a bacteriophage, whose real business in life is infecting the bacillus. It is even conceivable that the genetic information that enables the phage to induce the bacterium to produce a toxin was picked up elsewhere in the course of long intimacy with the animal host, and this may explain why the toxin is itself so closely similar to the host cell's own constituents.

It is certainly a strange relationship, without any of the straightforward predator-prey aspects that we used to assume for infectious disease. It is hard to see what the diphtheria bacillus has to gain in life from the capacity to produce such a toxin. Corynebacteria live well enough in the surface of human respiratory membranes, and the production of a necrotic pseudomembrane carries the risk of killing off the host and ending the relationship. It does not, in short, make much sense, and appears more like a biological mix-up than an evolutionary advantage.

The most malevolent of all microbial exotoxins for human beings is botulinus, and here there is no question as to the irrelevancy of the toxin. Tetanus and its toxin represent accidents in the same sense. It is interesting, though, that these organisms, like the diphtheria bacillus, and also the group A streptococcus and its erythrogenic toxin, are toxigenic because of having been infected by a phage. If it is a generality that bacteria will make exotoxins only when they are supplied with specifications by a virus, this is an extraordinary puzzle.

We were all reassured, when the first moon landing was ready to be made, that the greatest precautions would be taken to protect the life of the earth, especially human life, against infection by whatever there might be alive on the moon. And, in fact, the elaborate ceremony of lunar asepsis was performed after each of the early landings; the voyagers were masked and kept behind plate glass, quarantined away from contact with the earth until it was a certainty that we wouldn't catch something from them. The idea that germs are all around us, trying to get at us, to devour and destroy us, is so firmly rooted in modern consciousness that it made sense to think that strange germs, from the moon, would be even scarier and harder to handle.

It is true, of course, that germs are all around us; they comprise a fair proportion of the sheer bulk of the soil, and they abound in the air. But it is certainly not true that they are our natural enemies. Indeed, it comes as a surprise to realize that such a tiny minority of the bacterial populations of the earth has any interest at all in us. The commonest of encounters between bacteria and the higher forms of life take place after the death of the latter, in the course of recycling the elements of life. This is obviously the main business of the microbial world in general, and it has nothing to do with disease.

It is probably true that symbiotic relationships between bacteria and their metazoan hosts are much more common in nature than infectious disease, although I cannot prove this. But if you count up all the indispensable microbes that live in various intestinal tracts, supplying essential nutrients or providing enzymes for the breakdown of otherwise indigestible food, and add all the peculiar bacterial aggregates that live like necessary organs in the tissues of many insects, plus all the bacterial symbionts engaged in nitrogen fixation in collaboration with legumes, the total mass of symbiotic life is overwhelming. Alongside, the list of important bacterial infections of human beings is short indeed.

It might be different, I suppose, if we had learned less about sanitation, nutrition, and crowding, and it is in fact different for the newborn children in places where these things are not done well. The greatest cause of infant mortality, far and away, is enteric infection spread by a contaminated environment. But, by and large, infection has become a relatively minor threat to life as we have civilized ourselves and installed plumbing, and even less a threat now that we have antibiotics.

But even before all this, when times were uniformly awful everywhere, in the centuries of the great plagues, the war between microbes and men was never really an event of great scale, and more often than not, the violence of those diseases was due primarily to the violence of the host's defense mechanisms. Leprosy, like tuberculosis, is a highly destructive disease, but the destruction is in large part immunological, under governance of the host. The major lesions of syphilis, including those of arteritis and perhaps also tabes, are based, at least in part, on immunological reactions in response to the spirochete.

Today, with so much of infectious disease under control, we are left with a roster of important illnesses which it has become fashionable to call "degenerative." They include chronic diseases of the brain and cord, chronic nephritis, arthritic arteriosclerosis, and various disorders caused by impedance to blood circulation. Although the underlying mechanisms governing such diseases are still largely mysterious, it is becoming the popular view that many of them may be the result of environmental influences—the things we eat or breathe or touch. As in so much of the thinking about cancer, we are in search of outside causes for the things that go wrong.

It may turn out, however, when we have learned more about pathogenesis in general, that most of the events that underlie the tissue damage in these diseases are host mechanisms, under host control. We are vulnerable because of our very intricacy and complexity. We are systems of mechanisms, subject to all the small disturbances, tiny monkey wrenches, that can, in the end, produce the wracking and unhinging of interminable chains of coordinated, meticulously timed interaction.

On Natural Death

THERE ARE SO MANY new books about dying that there are now special shelves set aside for them in bookshops, along with the health-diet and home-repair paperbacks and the sex manuals. Some of them are so packed with detailed information and step-by-step instructions for performing the function that you'd think this was a new sort of skill which all of us are now required to learn. The strongest impression the casual reader gets, leafing through, is that proper dying has become an extraordinary, even an exotic experience, something only the specially trained get to do.

Also, you could be led to believe that we are the only creatures capable of the awareness of death, that when all the rest of nature is being cycled through dying, one generation after another, it is a different kind of process, done automatically and trivially, more "natural," as we say.

An elm in our backyard caught the blight this summer and dropped stone dead, leafless, almost overnight. One weekend it was a normal-looking elm, maybe a little bare in spots but nothing alarming, and the next weekend it was gone, passed over, departed, taken. Taken is right, for the tree surgeon came by yesterday with his crew of young helpers and their cherry picker, and took it down branch by branch and carted it off in the back of a red truck, everyone singing.

The dying of a field mouse, at the jaws of an amiable household cat, is a spectacle I have beheld many times. It used to make me wince. Early in life I gave up throwing sticks at the cat to make him drop the mouse, because the dropped mouse regularly went ahead and died anyway, but I always shouted unaffections at the cat to let him know the sort of animal he had become. Nature, I thought, was an abomination.

Recently I've done some thinking about that mouse, and I wonder if his dying is necessarily all that different from the passing of our elm. The main difference, if there is one, would be in the matter of pain. I do not believe that an elm tree has pain receptors, and even so, the blight seems to me a relatively painless way to go even if there were nerve endings in a tree, which there are not. But the mouse dangling tail-down from the teeth of a gray cat is something else again, with pain beyond bearing, you'd think, all over his small body.

There are now some plausible reasons for thinking it is

not like that at all, and you can make up an entirely different story about the mouse and his dying if you like. At the instant of being trapped and penetrated by teeth, peptide hormones are released by cells in the hypothalamus and the pituitary gland; instantly these substances, called endorphins, are attached to the surfaces of other cells responsible for pain perception; the hormones have the pharmacologic properties of opium; there is no pain. Thus it is that the mouse seems always to dangle so languidly from the jaws, lies there so quietly when dropped, dies of his injuries without a struggle. If a mouse could shrug, he'd shrug.

I do not know if this is true or not, nor do I know how to prove it if it is true. Maybe if you could get in there quickly enough and administer naloxone, a specific morphine antagonist, you could turn off the endorphins and observe the restoration of pain, but this is not something I would care to do or see. I think I will leave it there, as a good guess about the dying of a cat-chewed mouse, perhaps about dying in general.

Montaigne had a hunch about dying, based on his own close call in a riding accident. He was so badly injured as to be believed dead by his companions, and was carried home with lamentations, "all bloody, stained all over with the blood I had thrown up." He remembers the entire episode, despite having been "dead, for two full hours," with wonderment:

> It seemed to me that my life was hanging only by the tip of my lips. I closed my eyes in order, it seemed to me, to help push it out, and took pleasure in growing languid and letting

myself go. It was an idea that was only floating on the surface of my soul, as delicate and feeble as all the rest, but in truth not only free from distress but mingled with that sweet feeling that people have who have let themselves slide into sleep. I believe that this is the same state in which people find themselves whom we see fainting in the agony of death, and I maintain that we pity them without cause. . . . In order to get used to the idea of death, I find there is nothing like coming close to it.

Later, in another essay, Montaigne returns to it:

If you know not how to die, never trouble yourself; Nature will in a moment fully and sufficiently instruct you; she will exactly do that business for you; take you no care for it.

The worst accident I've ever seen was on Okinawa, in the early days of the invasion, when a jeep ran into a troop carrier and was crushed nearly flat. Inside were two young MPs, trapped in bent steel, both mortally hurt, with only their heads and shoulders visible. We had a conversation while people with the right tools were prying them free. Sorry about the accident, they said. No, they said, they felt fine. Is everyone else okay, one of them said. Well, the other one said, no hurry now. And then they died.

Pain is useful for avoidance, for getting away when there's time to get away, but when it is end game, and no way back, pain is likely to be turned off, and the mechanisms for this are wonderfully precise and quick. If I had to design an ecosystem in which creatures had to live off each other and in which dying was an indispensable part of living, I could not think of a better way to manage.

A Trip Abroad

I DO NOT BELIEVE for a minute that we are nearing the end of human surprise, despite resonantly put arguments by wonderfully informed scientists who tell us that after molecular biology and astrophysics there is really very little more to learn of substance. Except, they always add, for the nature of human consciousness, and that, they always add, is placed beyond our reach by the principle of indeterminacy; that is, our thought is so much at the center of life that it cannot sit still while we examine it.

But there may be a way out of this; it may turn out that consciousness is a much more generalized mechanism, shared round not only among ourselves but with all the other conjoined things of the biosphere. Thus, since we are not, perhaps, so absolutely central, we may be able to get a look at it, but we will need a new technology for this kind of neurobiology; in which case we will likely find that we have a whole eternity of astonishment stretching out ahead of us. Always assuming, of course, that we're still here.

We must rely on our scientists to help us find the way through the near distance, but for the longer stretch of the future we are dependent on the poets. We should learn to question them more closely, and listen more carefully. A poet is, after all, a sort of scientist, but engaged in a qualitative science in which nothing is measurable. He lives with data that cannot be numbered, and his experiments can be done only once. The information in a poem is, by definition, not reproducible. His pilot runs involve a recognition of things that pop into his head. The skill consists in his capacity to decide quickly which things to retain, which to eject. He becomes an equivalent of scientist, in the act of examining and sorting the things popping in, finding the marks of remote similarity, points of distant relationship, tiny irregularities that indicate that this one is really the same as that one over there only more important. Gauging the fit, he can meticulously place pieces of the universe together, in geometric configurations that are as beautiful and balanced as crystals. Musicians and painters listen, and copy down what they hear.

I wish that poets were able to give straight answers to straight questions, but that is like asking astrophysicists to make their calculations on their fingers, where we can watch the process. What I would like to know is: how should I feel about the earth, these days? Where has all the old nature gone? What became of the wild, writhing, unapproachable mass of the life of the world, and what happened to our old, panicky excitement about it? Just in fifty years, since I was a small boy in a suburban town, the world has become a structure of steel and plastic, intelligible and diminished. Mine was a puzzling maple grove of a village on the outskirts of New York City, and it vanished entirely, trees and all. It is now a syncytium of apartment houses, sprouting out of a matrix of cement flooded and jelled over an area that once contained 25,000 people who walked on grass. Now I live in another, more distant town, on a street with trees and lawns, and at night I can hear the soft sound of cement, moving like incoming tide, down the Sunrise Highway from New York.

If you fly around the earth and keep looking down, you will see that we have inserted ourselves everywhere. All fields are tilled. All mountains have been climbed and are being covered with concrete and plastic; some mountains, like the Appalachians, are simply cut down like trees. The fish are all trapped and domesticated, farmed in zoned undersea pastures. As for the animals, we will never have enough plastic bags for the bodies; soon the only survivors will be the cattle and sheep for the feeding of us, and the dogs and cats in our houses, fed while it lasts on the flesh of whales. And the rats and roaches, and a few reptiles.

The winged insects are vanishing, the calcium in the shells of eggs, and the birds.

We have dominated and overruled nature, and from now on the earth is ours, a kitchen garden until we learn to make our own chlorophyll and float it out in the sun inside plastic membranes. We will build Scarsdale on Mount Everest.

We will have everything under control, managed. Then what do we do? On long Sunday afternoons, what do we do, when there is nobody to talk to but ourselves?

It is because of these problems that we are now engaged in scrutinizing with such intensity the dark, bare flanks of Mars, hideous with lifelessness as it seems to be. We are like a family looking through travel brochures.

There is such a thing as too much of this. Because of our vast numbers and the rapidity with which we have developed prosthetic devices enabling us to hear and see each other, in person, all around the earth, we have become obsessed with ourselves. You'd think, to hear us think, that there was nothing else of significance on the earth except us.

Perhaps we should try to get away, for a while anyway. A change of scene might do us a world of good.

The trouble is, the barrenness of all the local planets. Perhaps we will be unlucky with our green thumbs, unable to create or maintain the faintest gasp of life on Mars or Titan. What's to stop us from looking elsewhere, farther on? If we can learn to navigate before the solar wind, we could, out there, hoist sail and tack our way out to where the wind fades off, practicing free-falls all the while, probing for gravity, trusting to luck, taking our chances. It would be like old times.

On Meddling

WHEN YOU ARE confronted by any complex social system, such as an urban center or a hamster, with things about it that you're dissatisfied with and anxious to fix, you cannot just step in and set about fixing with much hope of helping. This realization is one of the sore discouragements of our century. Jay Forrester has demonstrated it mathematically, with his computer models of cities in which he makes clear that whatever you propose to do, based on common sense, will almost inevitably make matters worse rather than better. You cannot meddle with one part of a complex system from the outside without the almost certain risk of setting off disastrous events that you hadn't counted on in other, remote parts. If you want to fix something you are first obliged to understand, in detail, the whole system, and for very large systems you can't do this without a very large computer. Even then, the safest course seems to be to stand by and wring hands, but not to touch.

Intervening is a way of causing trouble.

If this is true, it suggests a new approach to the problems of cities, from the point of view of experimental pathology: maybe some of the things that have gone wrong are the result of someone's efforts to be helpful.

It makes a much simpler kind of puzzle. Instead of trying to move in and change things around, try to reach in gingerly and simply extract the intervener.

The identification and extraction of isolated meddlers is the business of modern medicine, at least for the fixing of diseases caused by identifiable microorganisms. The analogy between a city undergoing disintegration and a diseased organism does not stretch the imagination too far. Take syphilis, for instance. In the old days of medicine, before the recognition of microbial disease mechanisms, a patient with advanced syphilis was a complex system gone wrong without any single, isolatable cause, and medicine's approach was, essentially, to meddle. The analogy becomes more spectacular if you begin imagining what would happen if we knew everything else about modern medicine with the single exception of microbial infection and the spirochete. We would be doing all sorts of things to intervene: inventing new modifications of group psychotherapy to correct the flawed thinking of general paresis, transplanting hearts with aortas attached for cardiovascular lues, administering immunosuppressant drugs to reverse the autoimmune reactions in tabes, enucleating gummas from the liver, that sort of effort. We might even be wondering about the role of stress in this peculiar, "multifactorial," chronic disease, and there would be all kinds of suggestions for "holistic" approaches, ranging from changes in the home environment to White House commissions on the role of air pollution. At an earlier time we would have been busy with bleeding, cupping, and purging, as indeed we once were. Or incantations, or shamanist fits of public ecstasy. Anything, in the hope of bringing about a change for the better in the whole body.

These were the classical examples of medical intervention in the prescientific days, and there can be no doubt that most of them did more harm than good, excepting perhaps the incantations.

With syphilis, of course, the problem now turns out to be simple. All you have to do, armed with the sure knowledge that the spirochete is the intervener, is to reach in carefully and eliminate this microorganism. If you do this quickly enough, before the whole system has been shaken to pieces, it will put itself right and the problem solves itself.

Things are undoubtedly more complicated in pathological social systems. There may be more than one meddler involved, maybe a whole host of them, maybe even a *system* of meddlers infiltrating all parts of the system you're trying to fix. If this is so, then the problem is that much harder, but it is still approachable, and soluble, once you've identified the fact of intervention.

It will be protested that I am setting up a new sort of straw demonology, postulating external causes for pathological events that are intrinsic. Is it not in the nature of complex social systems to go wrong, all by themselves, without external cause? Look at overpopulation. Look at Calhoun's famous model, those crowded colonies of rats and their malignant social pathology, all due to their own skewed behavior. Not at all, is my answer. All you have to do is find the meddler, in this case Professor Calhoun himself, and the system will put itself right. The trouble with those rats is not the innate tendency of crowded rats to go wrong, but the scientist who took them out of the world at large and put them into too small a box.

I do not know who the Calhouns of New York City may be, but it seems to me a modest enough proposal that they be looked for, identified, and then neatly lifted out. Without them and their intervening, the system will work nicely. Not perfectly, perhaps, but livably enough.

We have a roster of diseases which medicine calls "idiopathic," meaning that we do not know what causes them. The list is much shorter than it used to be; a century ago, common infections like typhus fever and tuberculous meningitis were classed as idiopathic illnesses. Originally, when it first came into the language of medicine, the term had a different, highly theoretical meaning. It was assumed that most human diseases were intrinsic, due to inbuilt failures of one sort or another, things gone wrong with various internal humors. The word "idiopathic" was intended to mean, literally, a disease having its own origin, a primary disease without any external cause. The list of such disorders has become progressively shorter as medical science has advanced, especially within this century, and the meaning of the term has lost its doctrinal flavor; we use "idiopathic" now to indicate simply that the cause of a particular disease is unknown. Very likely, before we are finished with medical science, and with luck, we will have found that all varieties of disease are the result of one or another sort of meddling, and there will be no more idiopathic illness.

With time, and a lot more luck, things could turn out this way for the social sciences as well.

On Committees

THE MARKS OF SELFNESS are laid out in our behavior irreversibly, unequivocally, whether we are assembled in groups or off on a stroll alone. Nobody can be aware of the unique immunologic labels of anyone else, outside a laboratory, nor can we smell with any reliability the pheromonal differences among ourselves. So, all we have to go by is how we walk, sound, write letters, turn our heads. We are infallible at this. Nobody is really quite like anyone else; there are reminders here and there, but no exact duplicates; we are four billion unique individuals.

Thus when committees gather, each member is necessarily an actor, uncontrollably acting out the part of himself, reading the lines that identify him, asserting his identity.

This takes quite a lot of time and energy, and while it is going on there is little chance of anything else getting done. Many committees have been appointed in one year and gone on working well into the next decade, with nothing much happening beyond these extended, uninterruptible displays by each member of his special behavioral marks.

If it were not for such compulsive behavior by the individuals, committees would be a marvelous invention for getting collective thinking done. But there it is. We are designed, coded, it seems, to place the highest priority on being individuals, and we must do this first, at whatever cost, even if it means disability for the group.

This is surely the driving idea behind democracy, and it is astonishing that the system works at all, let alone well. The individual is the real human treasure, and only when he has been cultivated to full expression of his selfness can he become of full value to society. Like many attractive social ideas, it is authentic, ancient Chinese. Integrity is the most personal of qualities; groups and societies cannot possess it until single mortals have it in hand. It is hard work for civilization.

But individuality can be carried too far, and you can see it happening almost all the time in committees. There are some very old words for criticizing the display of too much individuality. When someone becomes too separate, too removed, out of communication, his behavior is called egregious. This was once a nice word, meaning "out of the herd," signifying distinction and accomplishment, but by the linguistic process of pejoration the word took on an antisocial significance. Overindividuals are called peculiar, strange, eccentric. The worst sort are idiots, from *idios,* originally meaning personal and private.

These days, with the increasing complexity of the organizations in which we live and the great numbers of us becoming more densely packed together, the work of committees can be a deadly serious business. This is especially so when there is need to forecast the future. By instinct, each of us knows that this is a responsibility not to be trusted to any single person; we have to do it together.

Because of the urgency of the problems ahead, various modifications of the old standard committee have been devised in recent years, in efforts to achieve better grades of collective thought. There are the think tanks, hybrids between committee and factory, little corporations for thinking. There are governmental commissions and panels, made up of people brought to Washington and told to sit down together and think out collective thoughts. Industries have organized their own encounter groups, in which executives stride around crowded rooms bumping and shouting at each other in hopes of prodding out new ideas. But the old trouble persists: people assembled for group thought are still, first of all, individuals in need of expressing selfness.

The latest invention for getting round this is the Delphi technique. This was an invention of the 1960s, worked out by some RAND Corporation people dissatisfied with the way committees laid plans for the future. The method has a simple, almost silly sound. Instead of having meetings, questionnaires are circulated to the members of a group, and each person writes his answers out and sends them back, in silence. Then the answers are circulated to all members and they are asked to reconsider and fill out the questionnaires again, after paying attention to the other views. And so forth. Three cycles are usually enough. By that time as much of a consensus has been reached as can be reached, and the final answers are said to be substantially more reliable, and often more interesting, than first time around. In some versions, new questions can be introduced by the participants at the same time that they are providing answers.

It is almost humiliating to be told that Delphi works, sometimes wonderfully well. One's first reaction is resentment at still another example of social manipulation, social-science trickery, behavior control.

But, then, confronted by the considerable evidence that the technique really does work—at least for future-forecasting in industry and government—one is bound to look for the possibly good things about it.

Maybe, after all, this is a way of preserving the individual and all his selfness, and at the same time linking minds together so that a group can do collective figuring. The best of both worlds, in short.

What Delphi is, is a really quiet, thoughtful conversation, in which everyone gets a chance to *listen.* The background noise of small talk, and the recurrent sonic booms of vanity, are eliminated at the outset, and there is time to think. There are no voices, and therefore no rising voices. It is, when you look at it this way, a great discovery. Before Delphi, real listening in a committee meeting has always been a near impossibility. Each member's function was to talk, and while other people were talking the individual member was busy figuring out what he ought to say next in order to shore up his own original position. Debating is what committees really do, not thinking. Take away the need for winning points, leading the discussion, protecting one's face, gaining applause, shouting down opposition, scaring opponents, all that kind of noisy activity, and a group of bright people can get down to quiet thought. It is a nice idea, and I'm glad it works.

It is interesting that Delphi is the name chosen, obviously to suggest the oracular prophetic function served. The original Delphi was Apollo's place, and Apollo was the god of prophecy, but more than that. He was also the source of some of the best Greek values: moderation, sanity, care, attention to the rules, deliberation. Etymologically, in fact, Apollo may have had his start as a committee. The word *apollo* (and perhaps the related word *apello*) originally meant a political gathering. The importance of public meetings for figuring out what to do next must have been perceived very early as fundamental to human society, therefore needing incorporation into myth and the creation of an administrative deity; hence Apollo, the Dorian god of prophecy.

The Pythian prophetess of Delphi was not really supposed to enunciate clear answers to questions about the future. On the contrary, her pronouncements often contained as much vagueness as the *I Ching,* and were similarly designed to provide options among which choice was possible. She symbolized something more like the committee's agenda. When she collapsed in ecstasy on the tripod, murmuring ambiguities, she became today's questionnaire. The working out of the details involved a meticulous exegesis of the oracle's statements, and this was the task of the *exegetai,* a committee of citizens, partly elected by the citizens of Athens and partly appointed by the Delphi oracle. The system seems to have worked well enough for a long time, constructing the statutory and legal basis for Greek religion.

Today's Delphi thus represents a refinement of an ancient social device, with a novel modification of committee procedure constraining groups of people to think more quietly, and to listen. The method seems new, as a formal procedure, but it is really very old, perhaps as old as human society itself. For in real life, this is the way we've always arrived at decisions, even though it has always been done in a disorganized way. We pass the word around; we ponder how the case is put by different people; we read the poetry; we meditate over the literature; we play the music; we change our minds; we reach an understanding. Society evolves this way, not by shouting each other down, but by the unique capacity of unique, individual human beings to comprehend each other.

The Scrambler
in the Mind

LINGUISTIC SCHOLARS do a lot of arguing with each other over matters of theory, as they should. You would expect learned men who must spend their lives trying to figure out language, which is to say how to understand humanity all at once, to disagree often, even to become testy and impatient with each other, which, having more to worry about, they seem in fact to do rather more than their colleagues in other fields of academic science.

There is a hard technical problem which confronts linguists, and especially the savants who are today's version of philosophers. They are compelled to use as their sole research instrument the very apparatus that they wish to study, and this makes them especially vulnerable to the sort of hazard that physicists have had on their minds since Heisenberg. The closer linguists come to the center of their problem, the more they must manipulate the mechanism they are examining with precisely the same mechanism; no wonder that just at the moment when it seems within reach there is a twitch and a tremor and it shifts away in a blur.

You can see this happening in some of the books written about language, especially in books by writers who are not themselves professional and therefore cautious linguists. Brought in from outside to explain matters to the public at large, such writers always reach a point where suddenly the prose itself becomes wildly incomprehensible, disintegrating into nonsense. Usually this happens after the elements of linguistic logic have been nicely laid out, the fundamental notions of transformational grammar explained, the question whether some languages are more "complex" than others dealt with, and the mathematic techniques for deep analysis of sentences fully described. Equipped with so much powerfully usable information, the nonprofessional moves straight ahead, unaware that he is now stepping across the frontier into an unknown and maybe unknowable land, and he vanishes from sight.

I'm not sure what happens, exactly, at this stage of communication. I would like to believe that something goes wrong with the transmitter of information, and that what the reader is given is nonsense in the act of transmission, but I could be wrong about this. Perhaps it is plain, lucid prose after all, and the trouble is at my end, in my brain; maybe I do not possess receptors for this kind of talk.

Or maybe I alter it as it comes in, without realizing that I'm doing this.

I've had the same uncomfortable misgiving on other occasions, in matters not involving linguistics. Gödel's Theorem was once explained to me by a patient, gentle mathematician, and just as I was taking it all in, nodding appreciatively at the beauty of the whole idea, I suddenly felt something like the silent flicking of a mercury wall switch and it all turned to nonsense inside my head. I have had similar experiences listening to electronic music, and even worse ones reading poetry criticism. It is not like blanking out or losing interest or drifting off, not at all. My mind is, if anything, more alert, grasping avidly at every phrase, but then the switch is thrown and what comes in is transformed into an unfathomable code.

This brings me to my theory about the brain, my brain anyway. I believe there is a center someplace, maybe in the right hemisphere, which has a scrambling function similar to those electronic devices attached to the telephones of important statesmen which instantly convert all confidential sentences to gibberish.

Maybe there is a need for secrecy where language is involved. It is conceivable that if we had anything like full, conscious comprehension of what we are doing, our speech would be degraded to a permanent stammer or even into dead silence. It would be an impossible intellectual feat to turn out the simplest of sentences, the lovely Wallace Stevens sentence, for example: *"The man replied, Things as they are, are changed upon the blue guitar."* Doing that sort of thing, monitoring all the muscles, keeping an eye on the syntax, watching out for the semantic catastrophes risked by the slightest change in word order, taking care of the tone of voice and expression around the eyes and mouth, and worrying most of all about the danger of saying something meaningless, would be much harder to accomplish than if you were put in charge of your breathing and told to look after that function, breath by breath forever, with your conscious mind.

A scrambler in the brain would be a protective device, preserving the delicate center of the mechanism of language against tinkering and meddling, shielding the mind against information with which it has no intention of getting involved.

You might think that if there were a neuronal scrambler in one part of the brain, there ought to be a symmetrically placed descrambling center, somewhere in another lobule, capable of putting disintegrated information back into something like its original order. I doubt this. I concede that the brain is unlikely to risk filling itself up with totally meaningless noise, but I think it more likely that such really deep and dangerous scrambled notions as the true nature of speech are reassembled into unrecognizably pleasant experiences, like small talk or music or sleep. Some people, very quick on their feet, can catch a fleeting glimpse of a thought just at the moment of its disappearance into the scrambler, and poems by people like Stevens are made in this way. But for most of us the business is done automatically, out of sight, and I suppose this is just as well.

Notes on Punctuation

THERE ARE no precise rules about punctuation (Fowler lays out some general advice (as best he can under the complex circumstances of English prose (he points out, for example, that we possess only four stops (the comma, the semicolon, the colon and the period (the question mark and exclamation point are not, strictly speaking, stops; they are indicators of tone (oddly enough, the Greeks employed the semicolon for their question mark (it produces a strange sensation to read a Greek sentence which is a straightforward question: Why weepest thou; (instead of Why weepest thou? (and, of course, there are parentheses (which are surely a kind of punctuation making this whole matter much more complicated by having to count up the left-handed parentheses in order to be sure of closing with the right number (but if the parentheses were left out, with nothing to work with but the stops, we would have considerably more flexibility in the deploying of layers of meaning than if we tried to separate all the clauses by physical barriers (and in the latter case, while we might have more precision and exactitude for our meaning, we would lose the essential flavor of language, which is its wonderful ambiguity)))))))))))).

The commas are the most useful and usable of all the stops. It is highly important to put them in place as you go along. If you try to come back after doing a paragraph and stick them in the various spots that tempt you you will discover that they tend to swarm like minnows into all sorts of crevices whose existence you hadn't realized and before you know it the whole long sentence becomes immobilized and lashed up squirming in commas. Better to use them sparingly, and with affection, precisely when the need for each one arises, nicely, by itself.

I have grown fond of semicolons in recent years. The semicolon tells you that there is still some question about the preceding full sentence; something needs to be added; it reminds you sometimes of the Greek usage. It is almost always a greater pleasure to come across a semicolon than a period. The period tells you that that is that; if you didn't get all the meaning you wanted or expected, anyway you got all the writer intended to parcel out and now you have to move along. But with a semicolon there you get a pleasant little feeling of expectancy; there is more to come; read on; it will get clearer.

Colons are a lot less attractive, for several reasons: firstly, they give you the feeling of being rather ordered around, or at least having your nose pointed in a direction you might not be inclined to take if left to yourself, and, secondly, you suspect you're in for one of those sentences that will be labeling the points to be made: firstly, secondly and so forth, with the implication that you haven't sense enough to keep track of a sequence of notions without having them numbered. Also, many writers use this system loosely and incompletely, starting out with number one and number two as though counting off on their fingers but then going on and on without the succession of labels you've been led to expect, leaving you floundering about searching for the ninethly or seventeenthly that ought to be there but isn't.

Exclamation points are the most irritating of all. Look! they say, look at what I just said! How amazing is my thought! It is like being forced to watch someone else's small child jumping up and down crazily in the center of the living room shouting to attract attention. If a sentence really has something of importance to say, something quite remarkable, it doesn't need a mark to point it out. And if it is really, after all, a banal sentence needing more zing, the exclamation point simply emphasizes its banality!

Quotation marks should be used honestly and sparingly, when there is a genuine quotation at hand, and it is necessary to be very rigorous about the words enclosed by the marks. If something is to be quoted, the *exact* words must be used. If part of it must be left out because of space limitations, it is good manners to insert three dots to indicate the omission, but it is unethical to do this if it means connecting two thoughts which the original author did not intend to have tied together. Above all, quotation marks should not be used for ideas that you'd like to disown, things in the air so to speak. Nor should they be put in place around clichés; if you want to use a cliché you must take full responsibility for it yourself and not try to fob it off on anon., or on society. The most objectionable misuse of quotation marks, but one which illustrates the dangers of misuse in ordinary prose, is seen in advertising, especially in advertisements for small restaurants, for example "just around the corner," or "a good place to eat." No single, identifiable, citable person ever really said, for the record, "just around the corner," much less "a good place to eat," least likely of all for restaurants of the type that use this type of prose.

The dash is a handy device, informal and essentially playful, telling you that you're about to take off on a different tack but still in some way connected with the present course —only you have to remember that the dash is there, and either put a second dash at the end of the notion to let the reader know that he's back on course, or else end the sentence, as here, with a period.

The greatest danger in punctuation is for poetry. Here it is necessary to be as economical and parsimonious with commas and periods as with the words themselves, and any marks that seem to carry their own subtle meanings, like dashes and little rows of periods, even semicolons and question marks, should be left out altogether rather than inserted to clog up the thing with ambiguity. A single exclamation point in a poem, no matter what else the poem has to say, is enough to destroy the whole work.

The things I like best in T. S. Eliot's poetry, especially in the *Four Quartets,* are the semicolons. You cannot hear them, but they are there, laying out the connections between the images and the ideas. Sometimes you get a glimpse of a semicolon coming, a few lines farther on, and it is like climbing a steep path through woods and seeing a wooden bench just at a bend in the road ahead, a place where you can expect to sit for a moment, catching your breath.

Commas can't do this sort of thing; they can only tell you how the different parts of a complicated thought are to be fitted together, but you can't sit, not even take a breath, just because of a comma,

The Deacon's Masterpiece

THE BRIGHTEST and most optimistic of my presentiments about the future of human health always seem to arouse a curious mixture of resentment and dismay among some very intelligent listeners. It is as though I'd said something bad about the future. Actually, all I claim, partly on faith and partly from spotty but unmistakable bits of evidence out of the past century of biomedical science, is that mankind will someday be able to think his way around the finite list of major diseases that now close off life prematurely or cause prolonged incapacitation and pain. In short, we will someday be a disease-free species.

Except for gaining a precise insight into the nature of human consciousness (which may elude us for a very long time, perhaps forever), I cannot imagine any other limits to the profundity of our understanding of living things. It may happen within the next few centuries, maybe longer, but when it does it will bring along, inevitably, the most detailed sorts of explanations for human disease mechanisms. It is an article of faith with me that we will then know how to intervene directly, to turn them around or prevent them.

Something like this has already happened for most of the major infections. Even though we are still in a primitive, earliest stage in the emergence of biology, as compared, say, to physics, we have accomplished enough basic science to permit the development of specific antimicrobial antiserums and an impressive list of safe, rational viral vaccines. Within fifty years after the recognition of bacteria as pathogens we had classified them and learned enough of their metabolic intricacies so that the field was ready for antibiotics. In the years since the late 1940s the first great revolution in technology in all the long history of medicine has occurred, and infectious diseases that used to devastate whole families have now been almost forgotten.

Events moved rapidly in the field of infection, and this may have represented abnormally good luck. For some of the others—heart disease, cancer, stroke, the senile psychoses, diabetes, schizophrenia, emphysema, hypertension, arthritis, tropical parasitism, and the like—we may be in for a longer, more difficult pull, but maybe not. With the pace of research having increased so rapidly in the last two decades, and the remarkable new young brains enlisted for the work of biology, we could be in for surprises at almost any time. Anyway, sooner or later, they will all become nonmysteries, accountable and controllable.

These prospects seem to me exciting and heartening, and it is hard to face the mute, sidelong glances of disapproval that remarks along these lines usually generate. You'd think I'd announced an ultimate calamity.

The trouble comes from the automatic question, "Then what?" It is the general belief that we need our diseases— that they are natural parts of the human condition. It goes against nature to tamper and manipulate them out of existence, as I propose. "Then what?" What on earth will we die of? Are we to go on forever, disease-free, with nothing to occupy our minds but the passage of time? What are the biologists doing to us? How can you finish life honorably, and die honestly, without a disease?

This last is a very hard question, almost too hard to face, and therefore just the sort of question you should look around for a poem to answer, and there is one. It is "The Deacon's Masterpiece, or, the Wonderful 'One-Hoss Shay,' " by Oliver Wendell Holmes. On the surface, this piece of rather dreadful nineteenth-century doggerel seems to concern the disintegration of a well-made carriage, but inside the verse, giving it the staying power to hold on to our minds for over a full century, is a myth about human death.

Moreover, it is a myth for the modern mind. It used to be the common wisdom that the living body was a vulnerable, essentially ramshackle affair, always at risk of giving way at one point or another, too complicated to stay in one piece. These days, with what is being learned about cellular biology, especially the form and function of subcellular structures and their macromolecular components, and the absolutely flawless arrangements for drawing on solar energy for the needs of all kinds of cells, the most impressive aspect of life is its sheer, tough power. With this near view, it becomes a kind of horrifying surprise to realize that things can go wrong—that a disorder of one part can bring down the whole amazing system. Looked at this way, disease seems a violation of nature, an appalling mistake. There must be a better way to go.

Thus, a detailed anatomy of Holmes's carriage can be read as a metaphor for a live organism—or, for that matter, a cell:

> Now in building of chaises, I tell you what,
> There is always *somewhere* a weakest spot—
> In hub, tire, felloe, in spring or thill,
> In panel, or crossbar, or floor, or sill,
> In screw, bolt, thoroughbrace—lurking still. . . .
> And that's the reason, beyond a doubt,
> That a chaise *breaks down,* but doesn't *wear out.*

This was the nineteenth-century view of disease, and the source of our trouble today. It assumes that there is always, somewhere, a weakest part, as though foreordained. Without fundamental, localized flaws in the system, it might simply age away. As it is, it is doomed to break down prematurely, unless you can figure out how to find and fix the flawed item. Dr. Holmes, in the science of his day, saw little likelihood of this, but he did see, in his imagination, the possibility of sustained perfection. The Deacon is his central, Olympian Creator, symbolizing Nature, incapable of fumbling. What he designs is the perfect organism.

. . . so built that it *couldn'* break daown . . .
. . . "the weakes' place mus' stan' the strain;
'N' the way t' fix it, uz I maintain,
Is only jest
T' make that place uz strong uz the rest."

Then, the successive acts of creation, collectively miraculous, scriptural in tone:

. . . the strongest oak,
That couldn't be split nor bent nor broke . . .
He sent for lancewood to make the thills;
The crossbars were ash, from the straightest trees,
The panels of white-wood, that cuts like cheese,
But lasts like iron for things like these. . . .

Step and prop-iron, bolt and screw,
Spring, tire, axle, and linchpin too,
Steel of the finest, bright and blue;
Thoroughbrace bison-skin, thick and wide;
Boot, top, dasher, from tough old hide . . .
That was the way he "put her through."
"There!" said the Deacon, "naow she'll dew!"

And dew she did. The chaise lived, in fact, for a full,
unblemished hundred years of undiseased life, each perfect
part supported by all the rest. It was born from the Deacon's
hands in 1755, the year of the great Lisbon earthquake, and
it died on the earthquake centenary, to the hour, in 1855.

The death was the greatest marvel of all. Up to the last
minute, the final turn of the splendid wheels, the thing
worked perfectly. There was aging, of course, and Holmes

concedes this in his myth, but it was a respectable, decent, proper sort of aging:

> A general flavor of mild decay,
> But nothing local, as one may say.
> There couldn't be—for the Deacon's art
> Had made it so like in every part
> That there wasn't a chance for one to start.

And then, the hour of death:

> . . .the wheels were just as strong as the thills,
> And the floor was just as strong as the sills,
> And the panels just as strong as the floor. . . .
> And the back crossbar as strong as the fore . . .
> And yet, *as a whole,* it is past a doubt
> In another hour it will be *worn out!*

What a way to go!

> First of November, 'Fifty-five!
> This morning the parson takes a drive.
> Now, small boys, get out of the way!
> Here comes the wonderful one-hoss shay,
> Drawn by a rat-tailed, ewe-necked bay.
> "Huddup!" said the parson. Off went they.

And the death scene itself. No tears, no complaints, no listening closely for last words. No grief. Just, in the way of the world, total fulfillment. Listen:

> All at once the horse stood still,
> Close by the meet'n'-house on the hill.
> First a shiver, and then a thrill,

Then something decidedly like a spill—
And the parson was sitting upon a rock,
At half past nine by the meet'n'-house clock—

And, finally, the view of the remains:

What do you think the parson found,
When he got up and stared around?
The poor old chaise in a heap or mound,
As if it had been to the mill and ground! . . .
. . . it went to pieces all at once—
All at once, and nothing first—
Just as bubbles do when they burst.

My favorite line in all this is one packed with the most
abundant meaning, promising aging as an orderly, drying-
up process, terminated by the most natural of events: "As
if it had been to the mill and ground!"

This is, in high metaphor, what happens when a healthy
old creature, old man or old mayfly, dies. There is no outside evil force, nor any central flaw. The dying is built into
the system so that it can occur at once, at the end of a
preclocked, genetically determined allotment of living.
Centralization ceases, the forces that used to hold cells together are disrupted, the cells lose recognition of each
other, chemical signaling between cells comes to an end,
vessels become plugged by thrombi and disrupt their walls,
bacteria are allowed free access to tissues normally forbid-
den, organelles inside cells begin to break apart; nothing
holds together; it is the bursting of billions of bubbles, all
at once.

What a way to go!

How to Fix the
Premedical Curriculum

THE INFLUENCE of the modern medical school on liberal-arts education in this country over the last decade has been baleful and malign, nothing less. The admission policies of the medical schools are at the root of the trouble. If something is not done quickly to change these, all the joy of going to college will have been destroyed, not just for that growing majority of undergraduate students who draw breath only to become doctors, but for everyone else, all the students, and all the faculty as well.

The medical schools used to say they wanted applicants as broadly educated as possible, and they used to mean it. The first two years of medical school were given over entirely to the basic biomedical sciences, and almost all entering students got their first close glimpse of science in those years. Three chemistry courses, physics, and some sort of biology were all that were required from the colleges. Students were encouraged by the rhetoric of medical-school catalogues to major in such nonscience disciplines as history, English, philosophy. Not many did so; almost all premedical students in recent generations have had their majors in chemistry or biology. But anyway, they were authorized to spread around in other fields if they wished.

There is still some talk in medical deans' offices about the need for general culture, but nobody really means it, and certainly the premedical students don't believe it. They concentrate on science.

They concentrate on science with a fury, and they live for grades. If there are courses in the humanities that can be taken without risk to class standing they will line up for these, but they will not get into anything tough except science. The so-called social sciences have become extremely popular as stand-ins for traditional learning.

The atmosphere of the liberal-arts college is being poisoned by premedical students. It is not the fault of the students, who do not start out as a necessarily bad lot. They behave as they do in the firm belief that if they behave any otherwise they won't get into medical school.

I have a suggestion, requiring for its implementation the following announcement from the deans of all the medical schools: henceforth, any applicant who is self-labeled as a "premed," distinguishable by his course selection from his classmates, will have his dossier placed in the third stack of three. Membership in a "premedical society" will, by itself, be grounds for rejection. Any college possessing something called a "premedical curriculum," or maintaining offices for people called "premedical advisers," will be excluded from recognition by the medical schools.

Now as to grades and class standing. There is obviously no way of ignoring these as criteria for acceptance, but it is the grades *in general* that should be weighed. And, since so much of the medical-school curriculum is, or ought to be, narrowly concerned with biomedical science, more attention should be paid to the success of students in other, nonscience disciplines before they are admitted, in order to assure the scope of intellect needed for a physician's work.

Hence, if there are to be MCAT tests, the science part ought to be made the briefest, and weigh the least. A knowledge of literature and languages ought to be the major test, and the scariest. History should be tested, with rigor.

The best thing would be to get rid of the MCATs, once and for all, and rely instead, wholly, on the judgment of the college faculties.

You could do this if there were some central, core discipline, universal within the curricula of all the colleges, which could be used for evaluating the free range of a student's mind, his tenacity and resolve, his innate capacity for the understanding of human beings, and his affection for the human condition. For this purpose, I propose that classical Greek be restored as the centerpiece of undergraduate education. The loss of Homeric and Attic Greek from American college life was one of this century's disasters. Putting it back where it once was would quickly make up for the dispiriting impact which generations of spotty Greek in translation have inflicted on modern thought. The capacity to read Homer's language closely enough to sense the terrifying poetry in some of the lines could serve as a shrewd test for the qualities of mind and character needed in a physician.

If everyone had to master Greek, the college students aspiring to medical school would be placed on the same footing as everyone else, and their identifiability as a separate group would be blurred, to everyone's advantage. Moreover, the currently depressing drift on some campuses toward special courses for prelaw students, and even prebusiness students, might be inhibited before more damage is done.

Latin should be put back as well, but not if it is handled, as it ought to be, by the secondary schools. If Horace has been absorbed prior to college, so much for Latin. But Greek is a proper discipline for the college mind.

English, history, the literature of at least two foreign languages, and philosophy should come near the top of the list, just below Classics, as basic requirements, and applicants for medical school should be told that their grades in these courses will count more than anything else.

Students should know that if they take summer work as volunteers in the local community hospital, as ward aides or laboratory assistants, this will not necessarily be held against them, but neither will it help.

Finally, the colleges should have much more of a say about who goes on to medical school. If they know, as they should, the students who are generally bright and also respected, this judgment should carry the heaviest weight for admission. If they elect to use criteria other than numerical class standing for recommending applicants, this evaluation should hold.

The first and most obvious beneficiaries of this new policy would be the college students themselves. There would no longer be, anywhere where they could be recognized as a coherent group, the "premeds," that most detestable of all cliques eating away at the heart of the college. Next to benefit would be the college faculties, once again in possession of the destiny of their own curriculum, for better or worse. And next in line, but perhaps benefiting the most of all, are the basic-science faculties of the medical schools, who would once again be facing classrooms of students who are ready to be startled and excited by a totally new and unfamiliar body of knowledge, eager to learn, unpreoccupied by the notions of relevance that are paralyzing the minds of today's first-year medical students already so surfeited by science that they want to start practicing psychiatry in the first trimester of the first year.

Society would be the ultimate beneficiary. We could look forward to a generation of doctors who have learned as much as anyone can learn, in our colleges and universities, about how human beings have always lived out their lives. Over the bedrock of knowledge about our civilization, the medical schools could then construct as solid a structure of medical science as can be built, but the bedrock would always be there, holding everything else upright.

A Brief Historical Note
on Medical Economics

I HAD FORGOTTEN what things were like in the good old days of medicine, and how different. I knew, of course, that the science and technology have undergone changes of great magnitude, and doctors now can accomplish such cures and relief of disability as were beyond the imagination when I was young. But there is another difference, and I'd forgotten about this.

I found it the other day while glancing through the yearbook of my class at Harvard Medical School at the time of graduation, in 1937. Albert Coons was the editor of the book, which contains the usual photographs of faculty eminences and administrators, and a smaller picture of each member of my class with a brief biographical sketch, including a line for the graduate's career plans. Coons, incidentally, who spent his life in immunologic research, beginning with the discovery of the method for labeling antibodies with fluorescent dyes, known ever since as the Coons Technique, stated in the note under his picture that he intended to practice internal medicine, in the East. Matter of fact, almost all my classmates who subsequently went into careers of research and teaching were quite sure at the time of graduation that they would become practitioners.

I digress. What I wanted to say is that Coons, as editor, decided to do something more ambitious for the yearbook than simply record the class statistics, and prepared a long questionnaire which was sent to all the alumni of the medical school from the classes which had graduated ten, twenty, and thirty years earlier. I remember the discussion at the time of sending the questionnaire, and particularly the feeling we all had that we were sampling the very extremes of seniority: graduates of the classes of 1927 and 1917 seemed to us figures from a very remote past, and those of 1907 were as far away as Galen.

To everyone's surprise, 60 percent of the 265 alumni filled out the questionnaire and returned it, pretty good for amateur social scientists of that time.

The findings of greatest interest, presented in some detail in the yearbook, concerned the net incomes of the alumni, which were, by the standards of the day, significantly higher than the AMA's figures for American physicians in general. This was reassuring to my class. We knew that interns and residents got room and board but no salary to speak of. We were glad to hear that Harvard graduates did better financially once out in practice. We were of course quick to say to ourselves that it wasn't the money that mattered, just that it seemed fair to conclude that if they made all that much more money they were most likely finer physicians.

Now for the difference, and the surprise. The median net income of the group of 165 Harvard Medical School graduates, ten to thirty years out of school, was between $5,000 and $10,000 a year. In the ten-year class, 43 percent made less than $5,000. Only five men earned over $20,000, and a single surgeon, twenty years out, made $50,000. Seven graduates of the class of 1927 had incomes below $2,500.

The alumni were invited to send in comments along with the questionnaire, in a space marked "Remarks," with the understanding that since so much of the form was directed at finding out how much money they were making they might like to say something about life in general. As it turned out, most of the "Remarks" were also about money, a typical comment being the following: "I am satisfied with medicine as a life's work. However, I should recommend it only for the man who has plenty of money back of him. Many men never make much in medicine."

Forty-one years ago, that was the way it was.

Why Montaigne
Is Not a Bore

FOR THE WEEKEND TIMES when there is nothing new in the house to read, and it is raining, and nothing much to think about or write about, and the afternoon stretches ahead all bleak and empty, there is nothing like Montaigne to make things better.

He liked to scratch his ears: "Scratching is one of Nature's sweetest gratifications."

Writing skeptically about the accounts of miracles which were commonplace news items in his time, he remarks, "I have seen no more evident monstrosity and miracle in the world than myself. . . . The more I frequent myself and know myself, the more my deformity astonishes me, and the less I understand myself."

It is one of the encouraging aspects of our civilization that Montaigne has never gone out of print. Even in the first decades after his death, when he was politically disliked for having taken a middle-of-the-road stand between the two extremes then arguing over power, the essays went through four editions and had already been translated into English and Spanish. He has since been made available to all the written languages of the earth, and scholars of all nations have built prosperous careers on his three books.

I used to stumble my way through the Florio translation, hard going because of the antique language but worth all the trouble nonetheless, until the Donald Frame version in American English came out, and I was off and running. It became my habit to turn the top corner of any page on which something so remarkable was written that I knew I would want to find it again. I have a poor memory and need to do this sort of thing. Now, eight years or so later, more than half the corners are turned, so that the book looks twice as thick sitting on the table, and I have discovered a new interest in Montaigne: what is there on all those un-turned pages that I have read but forgotten, still there to be discovered?

He is resolved from the first page to tell you absolutely everything about himself, and so he does. At the greatest length, throughout all 876 pages of the Frame translation, he tells you and tells you about himself.

This ought to be, almost by definition, the achievement of a great bore. How does it happen that Montaigne is not ever, not on any of all those pages, even a bit of a bore? Not even in the interminable "Apology for Raymond Sebond," which I passed over as a dull essay for several years. I knew that he had translated a theological tract by Sebond to please his father, and the essay contained his thoughts following the tedious experience. So I passed it by at each reading, or leafed it through quickly, absorbing nothing, turning no corners. Then, one day, I got into it, and never got out again. Raymond Sebond is the least of concerns; having given a dutiful nod to his father and Sebond in the first paragraphs, and an obligatory homily on the usefulness of reason for arriving at truth, Montaigne simply turns his mind loose and writes whatever he feels like writing. Mostly, he wants to say that reason is not a special, unique gift of human beings, marking us off from the rest of Nature. Bees are better at organizing societies. Elephants are more concerned for the welfare of other elephants, and clever at figuring things out; they will fill up the man-dug elephant trap with timber and earth in order to bring the trapped elephant to the surface. He is not even sure that language is any more complex or subtle than the exchanges of gestures and fragrances among the beasts. He catalogues a long list of creatures, magpies, jackals, foxes, songbirds, horses, dogs, oxen, turtles, fish, lions, whatnot, in anecdotes drawn mostly from the ancient classics to show how reasonable and essentially amiable they are, demonstrating to his own satisfaction "how much superiority the animals have over us, and how feeble is our skill to imitate them." It is the greatest fun.

Montaigne makes friends in the first few pages of the book, and he becomes the best and closest of all your friends as the essays move along. To be sure, he does go on and on about himself, but that self turns out to be the reader's self as well. Moreover, he does not pose, ever. He likes himself, to be sure, but is never swept off his feet after the fashion of bores. He is fond of his mind, and affectionately entertained by everything in his head.

He is, of course, a moralist and, like all the greatest moralists, also a humorist. I cannot imagine anyone reading Montaigne carefully, paying attention, concentrating on what he has to say, without smiling most of the time.

It is the easiest of conversations with a very old friend. Long silences are permitted, encouraged. The text is interrupted on every page by quotations from classical writers, in the custom of the day, but these often serve as mere resting places needing little focusing.

You can move through the essays casually, if you like, glancing at the pages as though at the view of the lawn through the window, waiting for something of interest to turn up. And then, "By the way," he says, and now you lean forward in your chair, and he begins to tell you what it is like to be a human being.

Self-appraisal is Montaigne's occupation in life. Not self-preoccupation or self-obsession, almost never self-approval. At best, a sort of qualified self-satisfaction, a puzzled resolve to put up with the man inside. For Montaigne, the nearest and most engrossing item in all of Nature is Montaigne; not the dearest but the nearest and therefore the easiest to get to know.

He was fascinated by his own inconstancy, and came to believe that inconsistency is an identifying biological characteristic of human beings in general. "We are all patchwork," he says, "so shapeless and diverse in composition that each bit, each moment, plays its own game."

There were no psychiatrists around in his day, but if there had been Montaigne would have had cautionary advice for them: "It has often seemed to me that even good authors are wrong to insist on fashioning a consistent and solid fabric out of us. They choose one general characteristic, and go and arrange and interpret all of man's actions to fit their picture; and if they cannot twist them enough, they go and set them down to dissimulation. . . . Nothing is harder for me than to believe in men's consistency, nothing easier than to believe in their inconsistency." We become so many different people in ourselves, at so many different times, he declares, "There is as much difference between us and ourselves as between us and others." The matter is too complicated for analysis; he concedes that the effort can be made "to probe the inside and discover what springs set men in motion," but, he warns, "Since this is an arduous and hazardous undertaking, I wish fewer people would meddle with it." This, mind you, four hundred years ago.

He despairs of making any sense of himself. He writes, "All contradictions may be found in me . . . bashful, insolent; chaste, lascivious; talkative, taciturn; tough, delicate; clever, stupid; surly, affable; lying, truthful; learned, ignorant; liberal, miserly and prodigal: all this I see in myself to some extent according to how I turn. . . . I have nothing to say about myself absolutely, simply and solidly, without confusion and without mixture, or in one word."

Having discovered and faced all this, he is not in the least troubled by it. He accepts the limitations and infirmities of

himself, and of humanity, with equanimity, even exuberance. "There is nothing so beautiful and legitimate as to
play the man well and duly; nor any science so arduous as
to know how to live this life of ours well and naturally. And
of our maladies the most wild and barbarous is to despise
our being. . . . For my part, I love life and cultivate it."

And so, on he goes, page after page, giving away his
thoughts without allowing himself to be constrained by any
discipline of consistency. "The greatest thing of the world,"
he writes, "is for a man to know how to be his own." As
it turns out, contrary to his own predictions, what emerges
is all his own, all of a piece, intact and solid as any rock. He
is, as he says everywhere, an ordinary man. He persuades
you of his ordinariness on every page. You cannot help but
believe him in this; he is, above all else, an honest and
candid man. And here is the marvel of his book: if Montaigne is an ordinary man, then what an encouragement,
what a piece of work is, after all, an ordinary man! You
cannot help but hope.

On Thinking
About Thinking

AT ANY WAKING MOMENT the human head is filled alive with molecules of thought called notions. The mind is made up of dense clouds of these structures, flowing at random from place to place, bumping against each other and caroming away to bump again, leaving random, two-step tracks like the paths of Brownian movement. They are small round structures, featureless except for tiny projections that are made to fit and then lock onto certain other particles of thought possessing similar receptors. Much of the time nothing comes of the activity. The probability that one notion will encounter a matched one, fitting closely enough for docking, is at the outset vanishingly small.

But when the mind is heated a little, the movement

speeds up and there are more encounters. The probability is raised.

The receptors are branched and complex, with configurations that are wildly variable. For one notion to fit with another it is not required that the inner structure of either member be the same; it is only the outside signal that counts for docking. But when any two are locked together they become a very small memory. Their motion changes. Now, instead of drifting at random through the corridors of the mind, they move in straight lines, turning over and over, searching for other pairs. Docking and locking continue, pairs are coupled to pairs, and aggregates are formed. These have the look of live, purposeful organisms, hunting for new things to fit with, sniffing for matched receptors, turning things over, catching at everything. As they grow in size, anything that seems to fit, even loosely, is tried on, stuck on, hung from the surface wherever there is room. They become like sea creatures, decorated all over with other creatures living as symbionts.

At this stage of its development, each mass of conjoined, separate notions, remembering and searching at the same time, shifts into its own fixed orbit, swinging in long elliptical loops around the center of the mind, rotating slowly as it goes. Now it is an idea.

Sometimes a mass of particles is packed so solidly that it begins, by something like gravitational force, to attract to itself everything else in the mind. The center then fails to hold, everything becomes skewed, other aggregates yaw and wobble into new orbits around the new dense mass, and nothing can escape the attraction. It is then a black hole, the mind seems to vanish from sight, and sleep occurs.

This is not the normal course of events, however. Under proper circumstances, when all the orbiting structures are evenly balanced, there is harmony. New notions, formed by impulses arriving from the outside, drift through the atmosphere. They lock to each other, pair up, double and redouble, and then, when things are going well, are swept onto the surface of one or another of the great orbiting aggregates. When the force of attraction is not strong enough for attachment, the new notions may simply move into tiny orbits around the aggregated ideas. This is not yet thinking, but it is the last stage in preparation for a thought.

The process of sorting and selecting, when many aggregates are simultaneously in flight and the separate orbits are now arranged in shimmering membranes very close to each other, is like a complicated, meticulously ordered dance. New notions are flung from one elliptical path into the next, collide with unmatched surfaces and bound away, to be caught and held in place by masses at a distance.

Now the motion of all the structures, large and small, becomes patterned and ceaselessly motoric, like the *Brandenburgs*. The aggregates begin to send out streamers, plumes of thought, which touch and adhere. Sometimes, not often but sometimes, all the particles are organized in aggregates and all the aggregates are connected, and the mind becomes a single structure, motile now and capable of purposeful, directional movement. Now the hunt begins again, for something similar, with matching receptors, *outside*.

Counterpoint is but one aspect of the process of combination, separation, recall, and recombination. Dance is only one aspect of the movement. The darting forward to meet new pairs of notions, built into new aggregates, the orbiting and occasional soaring of massive aggregates out of orbit and off into other spaces, most of all the continual switching of solitary particles of thought from one orbit into the next, like electrons, up and down depending on the charges around and the masses involved, accomplished as though by accident but always adhering to laws—all these have the look of music. There is no other human experience they can remind one of.

I suggest, then, that we turn it around. Instead of using what we can guess at about the nature of thought to explain the nature of music, start over again. Begin with music and see what this can tell us about the sensation of thinking. Music is the effort we make to explain to ourselves how our brains work. We listen to Bach transfixed because this is listening to a human mind. *The Art of the Fugue* is not a special pattern of thinking, it is not thinking about any particular thing. The spelling out of Bach's name in the great, unfinished layers of fugue at the end is no more than a transient notion, something flashed across the mind. The whole piece is not about thinking about something, it is about thinking. If you want, as an experiment, to hear the whole mind working, all at once, put on *The St. Matthew Passion* and turn the volume up all the way. That is the sound of the whole central nervous system of human beings, all at once.

On Embryology

A SHORT WHILE AGO, in mid-1978, the newest astonishment in medicine, covering all the front pages, was the birth
of an English baby nine months after conception in a dish.
The older surprise, which should still be fazing us all, is that
a solitary sperm and a single egg can fuse and become a
human being under any circumstance, and that, however
implanted, a mere cluster of the progeny of this fused cell
affixed to the uterine wall will grow and differentiate into
eight pounds of baby; this has been going on under our eyes
for so long a time that we've gotten used to it; hence the
outcries of amazement at this really minor technical modification of the general procedure—nothing much, really, beyond relocating the beginning of the process from the

fallopian tube to a plastic container and, perhaps worth mentioning, the exclusion of the father from any role likely to add, with any justification, to his vanity.

There is, of course, talk now about extending the technology beyond the act of conception itself, and predictions are being made that the whole process of embryonic development, all nine months of it, will ultimately be conducted in elaborate plastic flasks. When this happens, as perhaps it will someday, it will be another surprise, with more headlines. Everyone will say how marvelously terrifying is the new power of science, and arguments over whether science should be stopped in its tracks will preoccupy senatorial subcommittees, with more headlines. Meanwhile, the sheer incredibility of the process itself, whether it occurs in the uterus or *in* some sort of *vitro,* will probably be overlooked as much as it is today.

For the real amazement, if you want to be amazed, is the process. You start out as a single cell derived from the coupling of a sperm and an egg, this divides into two, then four, then eight, and so on, and at a certain stage there emerges a single cell which will have as all its progeny the human brain. The mere existence of that cell should be one of the great astonishments of the earth. People ought to be walking around all day, all through their waking hours, calling to each other in endless wonderment, talking of nothing except that cell. It is an unbelievable thing, and yet there it is, popping neatly into its place amid the jumbled cells of every one of the several billion human embryos around the planet, just as if it were the easiest thing in the world to do.

If you like being surprised, there's the source. One cell is switched on to become the whole trillion-cell, massive apparatus for thinking and imagining and, for that matter, being surprised. All the information needed for learning to read and write, playing the piano, arguing before senatorial subcommittees, walking across a street through traffic, or the marvelous human act of putting out one hand and leaning against a tree, is contained in that first cell. All of grammar, all syntax, all arithmetic, all music.

It is not known how the switching on occurs. At the very beginning of an embryo, when it is still nothing more than a cluster of cells, all of this information and much more is latent inside every cell in the cluster. When the stem cell for the brain emerges, it could be that the special quality of brainness is simply switched on. But it could as well be that everything else, every other potential property, is switched off, so that this most specialized of all cells no longer has its precursors' option of being a thyroid or a liver or whatever, only a brain.

No one has the ghost of an idea how this works, and nothing else in life can ever be so puzzling. If anyone does succeed in explaining it, within my lifetime, I will charter a skywriting airplane, maybe a whole fleet of them, and send them aloft to write one great exclamation point after another, around the whole sky, until all my money runs out.

Medical Lessons from History

It is customary to place the date for the beginnings of modern medicine somewhere in the mid-1930s, with the entry of sulfonamides and penicillin into the pharmacopoeia, and it is usual to ascribe to these events the force of a revolution in medical practice. This is what things seemed like at the time. Medicine was upheaved, revolutionized indeed. Therapy had been discovered for great numbers of patients whose illnesses had previously been untreatable. Cures were now available. As we saw it then, it seemed a totally new world. Doctors could now *cure* disease, and this was astonishing, most of all to the doctors themselves.

It was, no doubt about it, a major occurrence in medicine,

and a triumph for biological science applied to medicine but perhaps not a revolution after all, looking back from this distance. For the real revolution in medicine, which set the stage for antibiotics and whatever else we have in the way of effective therapy today, had already occurred one hundred years before penicillin. It did not begin with the introduction of science into medicine. That came years later. Like a good many revolutions, this one began with the destruction of dogma. It was discovered, sometime in the 1830s, that the greater part of medicine was nonsense.

The history of medicine has never been a particularly attractive subject in medical education, and one reason for this is that it is so unrelievedly deplorable a story. For century after century, all the way into the remote millennia of its origins, medicine got along by sheer guesswork and the crudest sort of empiricism. It is hard to conceive of a less scientific enterprise among human endeavors. Virtually anything that could be thought up for the treatment of disease was tried out at one time or another, and, once tried, lasted decades or even centuries before being given up. It was, in retrospect, the most frivolous and irresponsible kind of human experimentation, based on nothing but trial and error, and usually resulting in precisely that sequence. Bleeding, purging, cupping, the administration of infusions of every known plant, solutions of every known metal, every conceivable diet including total fasting, most of these based on the weirdest imaginings about the cause of disease, concocted out of nothing but thin air—this was the heritage of medicine up until a little over a century ago. It is astounding that the profession survived so long, and got away with so much with so little outcry. Almost everyone seems to have been taken in. Evidently one had to be a born skeptic, like Montaigne, to see through the old nonsense; but even Montaigne, who wrote scathingly about the illnesses caused by doctoring centuries before Ivan Illich, had little effect. Most people were convinced of the magical powers of medicine and put up with it.

Then, sometime in the early nineteenth century, it was realized by a few of the leading figures in medicine that almost all of the complicated treatments then available for disease did not really work, and the suggestion was made by several courageous physicians, here and abroad, that most of them actually did more harm than good. Simultaneously, the surprising discovery was made that certain diseases were self-limited, got better by themselves, possessed, so to speak, a "natural history." It is hard for us now to imagine the magnitude of this discovery and its effect on the practice of medicine. The long habit of medicine, extending back into the distant past, had been to treat everything with something, and it was taken for granted that every disease demanded treatment and might in fact end fatally if not treated. In a sober essay written on this topic in 1876, Professor Edward H. Clarke of Harvard reviewed what he regarded as the major scientific accomplishment of medicine in the preceding fifty years, which consisted of studies proving that patients with typhoid and typhus fever could recover all by themselves, without medical intervention, and often did better for being untreated than when they received the bizarre herbs, heavy metals, and fomentations that were popular at that time. Delirium tremens, a disorder long believed to be fatal in all cases unless subjected to constant and aggressive medical intervention, was observed to subside by itself more readily in patients left untreated, with a substantially improved rate of survival.

Gradually, over the succeeding decades, the traditional therapeutic ritual of medicine was given up, and what came to be called the "art of medicine" emerged to take its place. In retrospect, this art was really the beginning of the science of medicine. It was based on meticulous, objective, even cool observations of sick people. From this endeavor we learned the details of the natural history of illness, so that, for example, it came to be understood that typhoid and typhus were really two entirely separate, unrelated disorders, with quite different causes. Accurate diagnosis became the central purpose and justification for medicine, and as the methods for diagnosis improved, accurate prognosis also became possible, so that patients and their families could be told not only the name of the illness but also, with some reliability, how it was most likely to turn out. By the time this century had begun, these were becoming generally accepted as the principal responsibilities of the physician. In addition, a new kind of much less ambitious and flamboyant therapy began to emerge, termed "supportive treatment" and consisting in large part of plain common sense: good nursing care, appropriate bed rest, a sensible diet, avoidance of traditional nostrums and patent medicine, and a measured degree of trust that nature, in taking its course, would very often bring things to a satisfactory conclusion.

The doctor became a considerably more useful and respected professional. For all his limitations, and despite his inability to do much in the way of preventing or terminating illness, he could be depended on to explain things, to relieve anxieties, and to be on hand. He was trusted as an adviser and guide in difficult times, including the time of dying.

Meanwhile, starting in the last decade of the nineteenth century, the basic science needed for a future science of medicine got under way. The role of bacteria and viruses in illness was discerned, and research on the details of this connection began in earnest. The major pathogenic organisms, most notably the tubercle bacillus and the syphilis spirochete, were recognized for what they were and did. By the late 1930s this research had already paid off; the techniques of active and passive immunization had been worked out for diphtheria, tetanus, lobar pneumonia, and a few other bacterial infections; the taxonomy of infectious disease had become an orderly discipline; and the time was ready for sulfanilamide, penicillin, streptomycin, and all the rest. But it needs emphasizing that it took about fifty years of concentrated effort in basic research to reach this level; if this research had not been done we could not have guessed that streptococci and pneumococci exist, and the search for antibiotics would have made no sense at all. Without the long, painstaking research on the tubercle bacillus, we would still be thinking that tuberculosis was due to night air and we would still be trying to cure it by sunlight.

At that time, after almost a century of modified skepticism about therapy amounting finally to near nihilism, we abruptly entered a new era in which, almost overnight, it became possible with antibiotics to cure outright some of the most common and lethal illnesses of human beings—lobar pneumonia, meningitis, typhoid, typhus, tuberculosis, septicemias of various types. Only the virus diseases lay beyond reach, and even some of these were shortly to come under control—as in poliomyelitis and measles—by new techniques for making vaccines.

These events were simply overwhelming when they occurred. I was a medical student at the time of sulfanilamide and penicillin, and I remember the earliest reaction of flat disbelief concerning such things. We had given up on therapy, a century earlier. With a few exceptions which we regarded as anomalies, such as vitamin B for pellagra, liver extract for pernicious anemia, and insulin for diabetes, we were educated to be skeptical about the treatment of disease. Miliary tuberculosis and subacute bacterial endocarditis were fatal in 100 percent of cases, and we were convinced that the course of master diseases like these could never be changed, not in our lifetime or in any other.

Overnight, we became optimists, enthusiasts. The realization that disease could be turned around by treatment, provided that one knew enough about the underlying mechanism, was a totally new idea just forty years ago.

Most people have forgotten about that time, or are too young to remember it, and tend now to take such things for granted. They were born knowing about antibiotics, or the drugs simply fell by luck into their laps. We need reminding, now more than ever, that the capacity of medicine to deal with infectious disease was not a lucky fluke, nor was it something that happened simply as the result of the passage of time. It was the direct outcome of many years of hard work, done by imaginative and skilled scientists, none of whom had the faintest idea that penicillin and streptomycin lay somewhere in the decades ahead. It was basic science of a very high order, storing up a great mass of interesting knowledge for its own sake, creating, so to speak, a bank of information, ready for drawing on when the time for intelligent use arrived.

For example, it took a great deal of time, and work, before it could be understood that there were such things as hemolytic streptococci, that there were more than forty different serological types of the principal streptococcal species responsible for human disease, and that some of these were responsible for rheumatic fever and valvular heart disease. The bacteriology and immunology had to be done first, over decades, and by the early 1930s the work had progressed just far enough so that the connection between streptococcal infection and rheumatic fever could be perceived.

Not until this information was at hand did it become a certainty that rheumatic fever could be prevented, and with it a large amount of the chief heart disease affecting young people, if only a way could be found to prevent streptococcal infection. Similarly, the identification of the role of pneumococci in lobar pneumonia, of brucellae in undulant fever, typhoid bacilli in typhoid fever, the meningococci in epidemic meningitis, required the sorting out and analysis of what seemed at the time an immensely complicated body of information. Most of the labor in infectious-disease laboratories went into work of this kind in the first third of this century. When it was finished, the scene was ready for antibiotics.

What was not realized then and is not fully realized even now was how difficult it would be to accomplish the same end for the other diseases of man. We still have heart disease, cancer, stroke, schizophrenia, arthritis, kidney failure, cirrhosis, and the degenerative diseases associated with aging. All told, there is a list of around twenty-five major afflictions of man in this country, and a still more formidable list of parasitic, viral, and nutritional diseases in the less developed countries of the world, which make up the unfinished agenda of modern biomedical science.

How does one make plans for science policy with such a list? The quick and easy way is to conclude that these diseases, not yet mastered, are simply beyond our grasp. The thing to do is to settle down with today's versions of science and technology, and make sure that our health-care system is equipped to do the best it can in an imperfect world. The trouble with this approach is that we cannot afford it. The costs are already too high, and they escalate higher each year. Moreover, the measures available are simply not good enough. We cannot go on indefinitely trying to cope with heart disease by open-heart surgery, carried out at formidable expense after the disease has run its destructive course. Nor can we postpone such issues by oversimplifying the problems, which is what we do, in my opinion, by attributing so much of today's chronic and disabling disease to the environment, or to wrong ways of living. The plain fact of the matter is that we do not know enough about the facts of the matter, and we should be more open about our ignorance.

At the same time, and this will have a paradoxical sound, there has never been a period in medicine when the future has looked so bright. There is within medicine, somewhere beneath the pessimism and discouragement resulting from the disarray of the health-care system and its stupendous cost, an undercurrent of almost outrageous optimism about what may lie ahead for the treatment of human disease if we can only keep learning. The scientists who do research on the cardiovascular system are entirely confident that they will soon be working close to the center of things, and they no longer regard the mechanisms of heart disease as impenetrable mysteries. The cancer scientists, for all their public disagreements about how best to organize their research, are in possession of insights into the intimate functioning of normal and neoplastic cells that were unimaginable a few years back. The eukaryotic cell, the cell with a true nucleus, has itself become a laboratory instrument almost as neat and handy as the bacterial cell became in the early 1950s, ready now to be used for elucidating the mechanisms by which genes are switched on or off as developing cells differentiate or, as in the case of cancer cells, dedifferentiate. The ways in which carcinogenic substances, or viruses, or other factors still unrecognized intervene in the regulation of cell behavior represent problems still unsolved, but the problems themselves now appear to be approachable; with what has been learned in the past decade, they can now be worked on.

The neurobiologists can do all sorts of things in their

investigation, and the brain is an organ different from what it seemed twenty-five years ago. Far from being an intricate but ultimately simplifiable mass of electronic circuitry governed by wiring diagrams, it now has the aspect of a fundamentally endocrine tissue, in which the essential reactions, the internal traffic of nerve impulses, are determined by biochemical activators and their suppressors. The technologies available for quantitative study of individual nerve cells are powerful and precise, and the work is now turning toward the functioning of collections of cells, the centers for visual and auditory perception and the like, because work at this level can now be done. It is difficult to think of problems that cannot be studied, ever. The matter of consciousness is argued over, naturally, as a candidate for perpetual unapproachability, but this has more the sound of a philosophical discussion. Nobody has the feeling any longer, as we used to believe, that we can never find out how the brain works.

The immunologists, the molecular biochemists, and the new generation of investigators obsessed with the structure and function of cell membranes have all discovered that they are really working together, along with the geneticists, on a common set of problems: how do cells and tissues become labeled for what they are, what are the forces that govern the orderly development and differentiation of tissues and organs, and how are errors in the process controlled?

There has never been a time like it, and I find it difficult to imagine that this tremendous surge of new information will terminate with nothing more than an understanding of how normal cells and tissues, and organisms, function. I regard it as a certainty that there will be uncovered, at the same time, detailed information concerning the mechanisms of disease.

The record of the past half century has established, I think, two general principles about human disease. First, it is necessary to know a great deal about underlying mechanisms before one can really act effectively; one had to know that the pneumococcus causes lobar pneumonia before one could begin thinking about antibiotics. One did not have to know all the details, not even how the pneumococcus does its damage to the lungs, but one had to know that it was there, and in charge.

Second, for every disease there is a single key mechanism that dominates all others. If one can find it, and then think one's way around it, one can control the disorder. This generalization is harder to prove, and arguable—it is more like a strong hunch than a scientific assertion—but I believe that the record thus far tends to support it. The most complicated, multicell, multitissue, and multiorgan diseases I know of are tertiary syphilis, chronic tuberculosis, and pernicious anemia. In each, there are at least five major organs and tissues involved, and each appears to be affected by a variety of environmental influences. Before they came under scientific appraisal each was thought to be what we now call a "multifactorial" disease, far too complex to allow for any single causative mechanism. And yet, when all the necessary facts were in, it was clear that by simply switching off one thing—the spirochete, the tubercle bacillus, or a single vitamin deficiency—the whole array of disordered and seemingly unrelated pathologic mechanisms could be switched off, at once.

I believe that a prospect something like this is the likelihood for the future of medicine. I have no doubt that there will turn out to be dozens of separate influences that can launch cancer, including all sorts of environmental carcinogens and very likely many sorts of virus, but I think there will turn out to be a single switch at the center of things, there for the finding. I think that schizophrenia will turn out to be a neurochemical disorder, with some central, single chemical event gone wrong. I think there is a single causative agent responsible for rheumatoid arthritis, which has not yet been found. I think that the central vascular abnormalities that launch coronary occlusion and stroke have not yet been glimpsed, but they are there, waiting to be switched on or off.

In short, I believe that the major diseases of human beings have become approachable biological puzzles, ultimately solvable. It follows from this that it is now possible to begin thinking about a human society relatively free of disease. This would surely have been an unthinkable notion a half century ago, and oddly enough it has a rather apocalyptic sound today. What will we do about dying, and about all that population, if such things were to come about? What can we die of, if not disease?

My response is that it would not make all that much difference. We would still age away and wear out, on about the same schedule as today, with the terminal event being more like the sudden disintegration and collapse all at once of Oliver Wendell Holmes's well-known one-hoss shay.

The main effect, almost pure benefit it seems to me, would be that we would not be beset and raddled by disease in the last decades of life, as most of us are today. We could become a healthy species, not all that different from the healthy stocks of domestic plants and animals that we already take for granted. Strokes, and senile dementia, and cancer, and arthritis are not natural aspects of the human condition, and we ought to rid ourselves of such impediments as quickly as we can.

There is another argument against this view of the future which needs comment. It is said that we are fundamentally fallible as organisms, prone to failure, and if we succeed in getting rid of one set of ailments there will always be other new diseases, now waiting out in the forest, ready to take their places. I do not know why this is said, for I can see no evidence that such a thing has ever happened. To be sure, we have a higher incidence of chronic illness among older people than we had in the early years of this century, but that is because more of us have survived to become older people. No new disease, so far as I know, has come in to take the place of diphtheria, or smallpox, or whooping cough, or poliomyelitis. Nature being inventive, we will probably always have the odd new illness turning up, but not in order to fill out some ordained, predestined quota of human maladies.

Indeed, the official public-health tables of morbidity and mortality seem to be telling us this sort of thing already, even though, in all our anxiety, we seem unwilling to accept the news. We have already become in the Western world, on the record, the healthiest society in the history of humankind. Compared with a century ago, when every family was obliged to count on losing members throughout the early years of life, we are in a new world. A death in a young family has become a rare and dreadful catastrophe, no longer a commonplace event. Our estimated life expectancy, collectively, is longer this year than ever before in history. Part of this general and gradual improvement in health and survival is thanks to sanitary engineering, better housing, and, probably, more affluence, but a substantial part is also attributable, in recent years, to biomedical science. We have not done badly at all, and having begun so well, I see no reason why we should not do even better in the future.

My argument about how to do this will come as no surprise. I say that we must continue doing biomedical research, on about the same scale and scope as in the past twenty years, with expansion and growth of the enterprise being dependent on where new leads seem to be taking us. It is an expensive undertaking, but still it is less than 3 percent of the total annual cost of today's health industry, which at last count was over $140 billion, and it is nothing like as expensive as trying to live with the halfway technologies we are obliged to depend on in medicine today; if we try to stay with these for the rest of the century the costs will go through the ionosphere.

But I should like to insert a qualification in this argument, which may be somewhat more of a surprise, coming from a doctor. I believe that the major research effort, and far and away the greatest investment for the future, must be in the broad area of basic biological science. Here and there, to be sure, there will be opportunities for productive applied science, comparable, say, to the making of polio vaccine or the devising of multidrug therapy for childhood leukemia, but these opportunities will not come often, nor can they be forced into existence before their time. The great need now, for the medicine of the future, is for more information at the most fundamental levels of the living process. We are nowhere near ready for large-scale programs of applied science in medicine, for we do not yet know enough.

Good applied science in medicine, as in physics, requires a high degree of certainty about the basic facts at hand, and especially about their meaning, and we have not yet reached this point for most of medicine. Nor can we predict at this stage, with much confidence, which particular items of new information, from which fields, are the likeliest to be relevant to particular disease problems. In this circumstance there has to be a certain amount of guessing, even gambling, and my own view is that the highest yield for the future will come from whatever fields are generating the most interesting, exciting, and surprising sorts of information, most of all, surprising.

It seems to me that the safest and most prudent of bets to lay money on is surprise. There is a very high probability that whatever astonishes us in biology today will turn out to be usable, and useful, tomorrow. This, I think, is the established record of science itself, over the past two hundred years, and we ought to have more confidence in the process. It worked this way for the beginnings of chemistry; we obtained electricity in this manner; using surprise as a guide, we progressed from Newtonian physics to electromagnetism, to quantum mechanics and contemporary geophysics and cosmology. In biology, evolution and genetics were the earliest big astonishments, but what has been going on in the past quarter century is simply flabbergasting. For medicine, the greatest surprises lie still ahead of us, but they are there, waiting to be discovered or stumbled over, sooner or later.

I am arguing this way from the most practical, down-to-earth, pragmatic point of view. This kind of science is most likely, in the real world, to lead to significant improvements in human health, and at low cost. This is a point worth further emphasis, by the way. When medicine has really succeeded brilliantly in technology, as in immunization, for example, or antibiotics, or nutrition, or endocrine-replacement therapy, so that the therapeutic measures can be directed straight at the underlying disease mechanism and are decisively effective, the cost is likely to be very low indeed. It is when our technologies have to be applied halfway along against the progress of disease, or must be brought in after the fact to shore up the loss of destroyed tissue, that health care becomes enormously expensive. The deeper our understanding of a disease mechanism, the greater are our chances of devising direct and decisive measures to prevent disease, or to turn it around before it is too late.

So much for the practical side of the argument. We need much more basic science for the future of human health, and I will leave the matter there.

But I have one last thing to say about biological science. Even if I should be wrong about some of these predictions, and it turns out that we can blunder our way into treating or preventing one disease or another without understanding the process (which I will not believe until it happens), and if we continue to invest in biological science anyway, we cannot lose. The Congress, in its wisdom, cannot lose. The public cannot lose.

Here is what I have in mind.

These ought to be the best of times for the human mind, but it is not so. All sorts of things seem to be turning out wrong, and the century seems to be slipping through our fingers here at the end, with almost all promises unfulfilled. I cannot begin to guess at all the causes of our cultural sadness, not even the most important ones, but I can think of one thing that is wrong with us and eats away at us: we do not know enough about ourselves. We are ignorant about how we work, about where we fit in, and most of all about the enormous, imponderable system of life in which we are embedded as working parts. We do not really understand nature, at all. We have come a long way indeed, but just enough to become conscious of our ignorance. It is not so bad a thing to be totally ignorant; the hard thing is to be partway along toward real knowledge, far enough to be aware of being ignorant. It is embarrassing and depressing, and it is one of our troubles today.

It is a new experience for all of us. Only two centuries ago we could explain everything about everything, out of pure reason, and now most of that elaborate and harmonious structure has come apart before our eyes. We are *dumb.*

This is, in a certain sense, a health problem after all. For as long as we are bewildered by the mystery of ourselves, and confused by the strangeness of our uncomfortable connection to all the rest of life, and dumbfounded by the inscrutability of our own minds, we cannot be said to be healthy animals in today's world.

We need to know more. To come to realize this is what this seemingly inconclusive century has been all about. We have discovered how to ask important questions, and now we really do need, as an urgent matter, for the sake of our civilization, to obtain some answers. We now know that we cannot do this any longer by searching our minds, for there is not enough there to search, nor can we find the truth by guessing at it or by making up stories for ourselves. We cannot stop where we are, stuck with today's level of understanding, nor can we go back. I do not see that we have a real choice in this, for I can see only the one way ahead. We need science, more and better science, not for its technology, not for leisure, not even for health or longevity, but for the hope of wisdom which our kind of culture must acquire for its survival.